MW00585463

Lean for the Process Industries

Dealing with Complexity

Lean for the Process Industries

Dealing with Complexity

PETER L. KING

CRC Press
Taylor & Francis Group
Boca Raton London New York

CRC Press is an imprint of the
Taylor & Francis Group, an **informa** business

A PRODUCTIVITY PRESS BOOK

Productivity Press
Taylor & Francis Group
270 Madison Avenue
New York, NY 10016

© 2009 by Taylor & Francis Group, LLC
Productivity Press is an imprint of Taylor & Francis Group, an Informa business

International Standard Book Number-13: 978-1-4200-7851-0 (Hardcover)

Library of Congress Cataloging-in-Publication Data

King, Peter L.
 Lean for the process industries : dealing with complexity / Peter L. King.
  p. cm.
 Includes bibliographical references and index.
 ISBN 978-1-4200-7851-0
 1. Manufacturing processes. 2. Process control. I. Title.

TS183.K56 2009
658.5--dc22   2008051281

Visit the Taylor & Francis Web site at
http://www.taylorandfrancis.com

and the Productivity Press Web site at
http://www.productivitypress.com

Contents

PART II   Seeing the Waste

PART III
Lean Tools Needing Little Modification

PART IV
Lean Tools Needing a Different Approach

Acknowledgments

This book is the product of almost twenty years of on-the-job research, experimentation, and developing an understanding of how lean principles can be successfully applied to industries outside the traditional lean strike zone in metal working, parts production, and assembly. Many of the people who guided me, mentored me, and shared in the journey toward understanding of lean applicability to our processes are individually recognized below; to all of them, I offer my heartfelt thanks. It has been a great ride so far, and as I embark on a new path to apply my skills, I am extremely grateful to the people who have enabled me to arrive at this point in my career.

I have wanted for several years to document, in some formal fashion, all that I have learned and am continuing to learn as I apply these concepts to an ever-increasing variety of manufacturing problems, but my day job always got in my way. Now that I have a greater degree of control of how I choose to spend my time, it has become a much higher priority in my life. And as I began to capture my experiences on paper, I became even more aware that I was describing the collective efforts of many, many colleagues, co-workers, and clients. If I were to list them all, it might approach the size of the Newark phone directory (probably not, but it feels that way; I am indebted to so many people), so I'll have to restrict myself to a partial list.

- To Jim Anderson, one of my first DuPont managers, who enabled two career course corrections, thus providing key stepping stones toward my arrival at my first lean assignment.
- To Bill Sheirich, Vinay Sohoni, Ray Barnard, and John Anderson, IBM consultants who taught me the magic of Continuous Flow Manufacturing and got me started on my lean journey, and introduced us to Six Sigma as it was being practiced by Motorola, circa 1988.
- To Wayne Smith, who took everything that IBM taught the DuPont team and built structure, discipline, and rigor around it.
- To Susan Schall, who taught me the value of an understanding of statistical principles, well before Six Sigma became popular.
- To Paul Veenema, colleague and mentor, whom I had the pleasure of working with and learning from in two of my various reincarnations.

- To the many people in DuPont businesses who provided me the opportunity to develop and apply adaptations of lean to their processes and supply chains, including John New, Larry Mlinac, Pete Ellefson, Tom Holmes, Ed Reiff, Anne Kraft, John Rees, Mark Weining, Maureen DeFeo, Natalia Duchini, Iris Welch, Portia Yarborough, Donna Copley, and Tom Carroll.
- To Bill Alzos, friend and mentor, who taught me about the importance of interpersonal relationships in getting work done.
- To Moe Richard, who served as my manager on two occasions and, in each case, offered me the opportunity to move in new directions and the encouragement to succeed.
- To Bill McCabe, friend, colleague, and mentor, who reinforced my courage to take one of the biggest steps in my life by providing a successful model to follow.
- To Ted Brown, friend, colleague, teacher, and boss, one of the strongest proponents of positive change with whom I have ever worked; he has the vision to see the change, the will and energy to do the work, and the political acumen to be successful at it more often than most.
- To Laura Colosi, a lean colleague who had the patience and diligence to read and edit most of the original manuscript. Her comments and suggestions were insightful; her encouragement was priceless!
- To DuPont colleagues Bennett Foster, Cris Leyson, Nick Mans, and Rob Pinchot, who offered suggestions, comments, and much-needed reinforcement for the value of this endeavor.
- Very special thanks must go out to Dan Fogelberg, who, sadly, passed away late in 2007. Music made the many hours spent at my keyboard much more pleasant, and Full Circle, Home Free, Souvenirs, and Nether Lands were among my most frequent companions. As I developed my process for writing the manuscript, listening to Fogelberg became a part of the plan.
- To my daughters, Jennifer and Courtney, for all their love, support, and encouragement.
- Finally, to my wife, Bonnie, who volunteered to take on the role of Operations Manager of Lean Dynamics so that I would have the time to create this work, for all her time, love, and encouragement. I appreciate especially her patience during those periods when I would zone into writer mode and cut off all contact with the outside world.

Introduction

Lean manufacturing principles have been very widely adopted by manufacturing companies over the past twenty-five years, under a variety of names and acronyms. Before "lean" began its rise to its current position as the term of choice, efforts based on the same concepts were being implemented as continuous flow manufacturing (CFM), world-class manufacturing (WCM), just-in-time (JIT), zero inventory production (ZIP), and a host of others.

The companies following these practices have seen such dramatic improvement in performance that lean has spread across entire supply chains, leading users to map their warehouse management and logistics processes to drive out waste in those operations. More recently, lean has been applied to business processes (forecasting and demand management, accounts receivable) and across the entire enterprise (new product development processes, human resources talent management processes, legal department patent application processes, and so on).

The one gap, the one area where lean adoption seems to be lagging, is in manufacturing operations in what are known as the process industries. This group includes companies that produce:

- Consumer goods, such as toothpaste and shampoo
- Food and beverage, canned goods, frozen foods, bottled condiments, breakfast cereals, soft drinks
- House and automotive paints
- Synthetic fibers, and the products made from them, such as apparel and carpets
- Pulp and paper
- Glass and ceramics
- Base metals
- Plastics
- Fertilizers and crop protection products
- Batch chemicals, lubricants, adhesives
- Pharmaceuticals
- Petroleum

Books and magazines are full of examples and case studies of lean application to parts manufacture and assembly operations, such as automobiles, motorcycles, computers, medical instruments, consumer electronics, and many others of that type, but the coverage of application to the process industries has been sparse.

Some have concluded that, since a goal of lean is to achieve continuous flow, and since material is already flowing continuously in chemical and food plants, there would be no benefit from lean. There are two major fallacies in this point of view:

- Material is not continuously flowing in most of these plants. Although there may be some degree of continuous processing, more of it can be batch processing, and there is typically a lot of hold up with material stored in tanks or silos.
- Even in completely continuous processes, there is still very significant waste and opportunity for lean improvement.

Others have concluded that while there may be tremendous benefit in lean application, it just doesn't fit. In other words, their processes are just too different from traditional parts manufacture and assembly, thus making lean irrelevant to process plants. The processes are indeed very different, with different flow dynamics and behavior, so lean must be viewed from a somewhat different perspective, and tools and techniques adapted, in order to be properly applied to process plants.

Thankfully, a few companies in the process industries, including my former employer, have been successfully applying lean tools to their processes for almost twenty years (even before the term "lean production" was coined!) and have developed a track record of effective, practical, sustainable lean improvement.

Hence the need for and the purpose of this book: to document and explain the approaches and techniques that we have developed to facilitate lean applications in the process industries.

TECHNOLOGY AND CULTURE

The main focus of the book is on the areas where the process industries are indeed different, areas where a somewhat different view is needed to see waste and lean opportunities within the process, or areas where the tools required for solution must be adapted to be feasible for these processes. Areas of similarity between process plants and assembly operations are discussed only briefly, and only to paint a complete picture.

It has been said that 80 percent of lean is about the culture, and 20 percent about the technology. Although I agree with this view, you may get the opposite impression. This book may appear to be 80 percent on the technology and only about 20 percent on the culture. To some extent, that is in keeping with the purpose of the book, which is to focus on the differences in the two types of manufacturing industry. I believe that the methods used to fully engage all who participate in the process—to tap into their creativity, to make them feel a part of the process and of the organization, and to achieve true continuous improvement—transcend the type of process or the type of industry involved. Nonetheless, there is a significant amount of guidance herein about the cultural side of lean, but it is embedded in the "how to" portions of the book; for example, how to create a process industry value stream map (VSM), how to develop virtual cells, how to level production, how to implement pull. Therefore, that material may not be obvious as cultural guidance. But the implicit emphasis on culture is intentional: This intertwining of technical concepts with cultural requirements for successful implementation is both appropriate and synergistic; trying to transform a culture without some business purpose to drive it and without some processes or practices to be transformed rarely succeeds.

HOW THIS BOOK IS ORGANIZED

The book consists of four parts:

Part I: Lean and the Process Industries
- Chapter 1 is a brief overview of lean that puts the material in this book in relevant context. In case you're not completely familiar with lean, the essential elements are covered here for completeness.
- Chapter 2 describes the process industries and the characteristics that make lean application challenging. It explains differences in material flow patterns and dynamics that require adaptations of traditional lean approaches.

- Chapter 3 deals with the eight wastes commonly described in lean books, and how they manifest themselves in process operations. The root causes of some of these wastes that may be different in the process industries are explained.

Part II: Seeing the Waste

- In order to describe a process operation completely enough that both the waste and the primary causes of waste become apparent, the standard VSM must include additional parameters. Chapter 4 describes VSM requirements, both the standards and the additions.
- Chapter 5 describes how to read and analyze a process VSM, what can be learned from the material flow and the information flow portion of the map, and how and when to develop the future state VSM.

Part III: Lean Tools Needing Little Modification

Some of the more basic lean tools can be used in process operations in much the same way that they are in assembly operations. However, there are some additional considerations with process plants, which are explained in the following chapters:

- Total productive maintenance (TPM), and why TPM is even more important to process plants is explained in Chapter 6.
- The use of SMED techniques to reduce changeover time and losses is described in Chapter 7. The unique characteristics and challenges often encountered in process changeovers are discussed.
- Chapter 8 describes visual management techniques, with process plant examples.
- Kaizen events are described in Chapter 9, as well as an explanation as to why kaizens in process operations often require additional planning.

Part IV: Lean Tools Needing a Different Approach

Manufacturing processes have traditionally been described in terms of a volume variety continuum, with high-volume, low-variety production (such as automotive gasoline) at one end and high-variety, low-volume production (such as custom cabinet making) at the other. The markets served by the process industries have evolved to a situation where producers must satisfy both high volumes and high variety. A synthetic fiber plant, for example, may produce 300 million pounds of fiber each year, in several hundred individual end items or SKUs. And, in addition to high volume and high variety, process industries now have to deal with a third V: high demand variability.

The chapters in this part describe the techniques that have been developed to apply lean tools in a way that effectively deals with all of this complexity.

- Chapter 10 addresses the kinds of bottlenecks found in process lines, along with how to recognize them, how to determine root cause, and how to manage them.
- Cellular manufacturing, a lean concept often neglected in process plants, is explained in Chapter 11. You find out why it can be so beneficial to these industries and how to overcome perceived barriers.
- Chapter 12 deals with production leveling, heijunka, and a technique unique to the process industries sometimes called product wheels.
- In many process plants, make-to-order is not a possibility, but make-to-stock requires large inventories, due to the high product variety and variability. Finish-to-order is a reasonable and beneficial compromise, and is described in Chapter 13.
- Chapter 14 deals with pull replenishment systems, the unique challenges imposed by process equipment, and how to respond to them.
- When implementing pull, calculating required supermarket sizes is critical to smooth performance, and this concept is detailed in Chapter 15.
- The book concludes in Chapter 16 with a discussion of the role of management in lean, and the criticality of having effective business processes to guide lean operations on an ongoing basis. Actually, these processes are required by any operation, be it in parts assembly or in materials processing, so they may seem to fall outside the scope of this book. However, they are so vital for lean success that they must be mentioned.

MATHEMATICAL DETAIL

Recognizing that you might be an industrial engineer or manufacturing engineer, I felt that the book should include enough of the specific calculation methods to provide you the tools needed to apply the concepts described. On the other hand, you may just want to understand the concepts at a general level, so the intent has been to describe concepts in a way that they can be understood without digging into the specific equations. I hope this balance has been achieved.

If the solutions and practices described in this book seem complex, rest assured that they are, in fact, relatively simple and straightforward. It is the processes to which they are being applied that introduce or raise the level of complexity. The reality is that in applying even simple, straightforward tools to complex processes, the solutions appear complex; in reality, they are simple if applied using a disciplined, step-by-step methodology as described in this book. In short, the lean concepts explained here provide a practical way to deal with complexity.

RELATED TOPICS

Three additional topics (supply chain mapping, Six Sigma, and enterprise resource planning) are closely related to the material I present in this book, and are important to lean implementation. Because this book discusses the areas in which lean must be approached differently in the process industries, however, and these additional topics transcend the kind of industry to which they applied, I felt they were beyond the scope of my book, except for a brief overview here.

Supply Chain Mapping

The focus of this book is on production processes, but many of the same principles apply to supply chain analysis and improvement. Supply chain mapping (SCM) can bring the same understanding of flow dynamics in that environment as a VSM does in manufacturing. Many of the same wastes are found in inventory, transportation, defects, and extra processing. My colleagues and I have found that the information flow component of the map is even more important on SCMs than VSMs, because that is where the majority of root causes of supply chain waste begin.

Six Sigma

There has been a proliferation of books, articles, and presentations on the combination of lean and Six Sigma, tagged with names like Lean Sigma and Lean Six Sigma. This book does not go into a great deal of detail on Six Sigma; suffice it to say that many of the examples described were implemented under DMAIC. You may have already discovered that projects executed using the DMAIC framework are better defined, better executed, and far more sustainable than those not following such a disciplined approach. So if an operation has a Six Sigma focus, all major lean work should be executed as Six Sigma projects, or at least guided by the DMAIC framework. VSMs can be developed as part of a baseline project. Kaizen events are more effectively done if they are Green Belt projects.

There is an impression that Six Sigma will encumber, drag down, or lengthen a program. This is not true if done with common sense. You shouldn't shoot for three decimal place accuracy in all of your data; instead, accept what you can gather in a timely manner. Be selective about the use of the seventeen deliverables, but eliminate any of them by conscious choice, not by neglect. If done appropriately (that is, with a strong dose of common sense) the DMAIC framework can bring a beneficial degree of structure to lean projects without being overburdening, without adding waste. Anything in life can be done well, and anything can be done poorly. Six Sigma is no exception, and when done well can be an enabler to lean.

Enterprise Resource Planning

Information technology (IT) and enterprise resource planning (ERP) systems, while very important in lean implementation, are not discussed in this book. There are two reasons for this:

- The subject is complex and cannot be adequately covered in this text without overwhelming the main purpose.
- Many of the IT issues encountered in lean transcend industry type, so a detailed discussion would be outside the scope of this work.

Your company may have an IT department responsible for ERP implementation and support. If so, collaborate with it to decide how to configure those systems to incorporate the techniques described in this book.

Part I

Lean and the Process Industries

1

Lean Overview: Principles and Tools

In the mid-1980s, many companies in the Western Hemisphere began to adopt what was for them a new and different set of manufacturing principles. These went by various names: world-class manufacturing (WCM); just-in-time (JIT); zero inventory production (ZIP); stockless production; continuous flow manufacturing (CFM); and a myriad of others. All of these were based to a large extent on manufacturing principles and work processes developed by Toyota, beginning in the late 1940s, and continuing well into the 1970s and 1980s.

ORIGINS OF LEAN

No one had paid much attention to what Toyota and other Japanese manufacturers were doing until about 1980, when it became apparent that worldwide competitiveness was shifting dramatically. Not only was there intense competition from producers in the Far East, but entire industries that had been dominated by western manufacturers were seeing their markets taken over by Japanese manufacturers.

There were a number of beliefs about why this was happening: that the Japanese had adopted robotics and manufacturing automation to a much greater extent than their western counterparts, that Japanese wages were much lower, that Japanese manufacturers were dumping products into western markets below cost, that unions and labor laws were constricting western manufacturers, and a host of others. To get to the truth, westerners, notably MIT through the International Motor Vehicle Program and Richard Schonberger, began to take a closer look at what was really behind the competitive shift. These studies revealed that, although there may have been a slight degree of truth to these beliefs, they were largely myths and that there was something much more profound at work. They learned that it was not about robotics and automation, not about labor laws, but about a fundamentally different approach to manufacturing, with fundamentally different attitudes about quality, about waste, about the customer, about the role of the worker, about what was truly important in producing a product. They also discovered that the Japanese manufacturers had developed work processes and practices to actualize their attitudes and beliefs, and had begun to create a new work culture.

It seemed that history and geography played a large role in the differences between Japan and the western world that these mid-1980s studies found. Western manufacturing had been influenced by the two predominant manufacturing systems of the past three hundred years: craft production and mass production.

Craft production had been characterized by hand assembly of parts with little degree of interchangeability. Gauging systems of the 1700s were not precise, so that what was ⅛ inch to one fabricator might be different from what was ⅛ inch to another. Further, tooling was not capable of cutting hardened metals, so parts were typically machined and then hardened, and could change dimensionally, warp, or distort in the hardening process. As a consequence, no two parts were identical, so assembly required a skilled craftsman to grind, file, and polish parts so that they would fit together and function smoothly. So even though the completed product might function properly, it was slightly different from the one made before and the one following it. And the most valuable resource in the entire manufacturing process was the fitter, the craftsman doing the assembly.

All of this began to change in the late 1700s. Metal-working technology, tooling, and measurement systems improved to the point that interchangeable parts became a possibility. Eli Whitney, best known as the inventor of the cotton gin, was among the first to exploit this new capability. From 1801 to 1809, he manufactured 10,000 firearms for the U.S. Army using interchangeable parts, thus opening the door for mass production.

By 1900, mass production using interchangeable parts had become common. The next step-change in manufacturing systems came in 1913 when Henry Ford turned on the world's first moving assembly line at his Highland Park plant. This enabled complex mechanical products to be assembled at high speed and high throughput, and at far higher labor productivity than had been possible before. Over the next few decades, the assembly line became the standard for mass production.

The manufacturing studies conducted in the 1980s revealed that while mass production based on Ford's developments had become the worldwide standard, there were significant regional differences in philosophy and priority. In the United States, the post–World War II economy had created huge demand for consumer goods, so that although slogans may have said that "Quality is Job 1," making the day's production goal and shipping the product was likely the real "Job 1." Production volume appeared to be more important than quality. If quality wasn't perfect, it was covered by product warranties, and in the case of motor vehicles, by extensive networks of dealer repair shops. Correcting defects could be handled by "Mr. Goodwrench."

In Europe, even in 1980 when mass production had been a reality for sixty years, there were still strong cultural remnants of craft production. There was such a strong value for craftsmanship that some manufacturers felt that it was better to have skilled craftsmen at the end of the production line performing extensive rework rather than to build quality in. The MIT study found rework areas where "a third of the total effort in assembly occurred in this area. In other words, the German plant was expending more effort to fix the problems it had just created than the Japanese plant required to make a nearly perfect car the first time."

The environment that manufacturers in Japan found themselves in after World War II was understandably dramatically different from their western competitors. In addition to emerging from the war with a shattered economy and little capital to invest, Japanese manufacturers had other challenges:

- Japan is a relatively small country, where space is at a premium.
- There is not the abundance of natural resources found in the United States.
- The strength of labor laws and labor unions required better working conditions.
- They faced strong global competitors, eager to establish a foothold in the Japanese market.
- There was no immigrant workforce willing to do menial, highly repetitive work.
- Japan provided a relatively small customer base; Toyota, for example, made fewer cars in a year than Ford did in a day.
- There was a need for flexibility to meet a wide variety of vehicle needs.

Thus, the situation in Japan required manufacturers to do things very differently if they were to compete successfully. Toyota realized this perhaps more than other Asian manufacturers and, through the genius of Taiichi Ohno, Shigeo Shingo, Eiji Toyoda, and others, was able to create what has later become recognized as the most significant manufacturing advancement since Ford's moving assembly line.

They decided that because they had few resources available to them, they could not afford to waste any of them, so they embarked on a relentless pursuit of elimination of all waste. From Ford they learned to respect speed and flow. (Ford once remarked that although people flocked to his factories to see the automation, to see the conveyors, what he hoped they would see was the *flow*. Ford realized that although his degree of automation was impressive for its time, it was just a means to reach the real goal: flow.)

In fact, Toyoda and Ohno visited Ford's plants on several occasions, were impressed with what they saw and would have liked to have copied it, but were forced to make substantial modifications while adapting what they could. Toyota also adopted W. Edwards Deming's concept of total quality management (TQM), and the scientific approach for problem solving commonly referred to as the Deming cycle or plan-do-check-act (PDCA), as well as many ideas from the U.S. post-war training-within-industry (TWI) programs, such as continuous training, continuous improvement, and democratic management.

TPS BECOMES THE NEW PRODUCTION PARADIGM

What they began in the late 1940s, and further developed over the next twenty-five to thirty years, is now known as the Toyota Production System, or simply, TPS. So by the 1980s, as people began to take notice of what Toyota and other Japanese manufacturers were doing, a variety of manufacturing improvement methodologies sprang up, all loosely or directly based on Toyota's developments, but each with its own unique twists and terminology. That began to change in 1990 with the publication of *The Machine That Changed the World*, a description of the International Motor Vehicle Program, the five-year MIT study of automobile manufacturing around the world. A key thrust of the book was that the concepts and methodologies developed by Toyota were effective and powerful, and were giving Toyota and other Japanese counterparts significant competitive advantage. They coined the term *lean production* to represent TPS.

The suggestion to call this method of production lean production has been attributed to John Krafcik, the IMVP plant survey leader, who chose lean because TPS "uses less of everything compared with mass production." The word "lean" is appropriate when one thinks about the use in connection with performance athletes, often described as lean. Athletes engaged in training regimens are continuously reducing body fat, always honing muscles, particularly those most important to performance, just as lean is continuously reducing waste while improving processes, particularly those most important to flow and to serving customers. Cyclists and runners are described as flexible, nimble, agile, and fast, attributes found in lean manufacturing facilities. Successful athletes are focused on removing excess fat rather than all tissue, as companies embodying true lean principles are focused on removing waste rather than all costs. Most importantly, to become a successful athletic competitor requires dedication, commitment, and discipline; athletes realize that getting into competitive condition is not a one-time task, but something that requires ongoing discipline to maintain performance and to continue to improve. And so it is with manufacturing; becoming lean is not a one-time series of kaizen events, but an institutionalization of a continuous improvement mind-set within the company's culture.

Since its introduction in 1990 in *The Machine That Changed the World*, lean has become the universal term of choice to represent TPS and the principles and practices that have enabled Toyota to become a worldwide leader in automobile production.

The term *lean thinking* has been used to describe the overarching philosophies that drove Toyota to develop TPS, and now drive companies who are successfully becoming lean. So what is lean thinking? It is a company-wide continuous improvement approach to eliminate waste and add value for your customer. The book *Lean Thinking* by James Womack and Daniel Jones identified five interlocking principles: (1) specify the *value*, (2) identify the *value stream*, (3) make the value *flow* without interruption, (4) let the customer *pull* value from the producer, and (5) pursue *perfection*. Table 1.1 provides a brief definition of the five principles, which also incorporate ideas in *The Toyota Way* by Jeffrey Liker, in *Easier, Simpler, Faster* by Jean Cunningham and Duane Jones, and in other Productivity Press publications, including the *Shopfloor* Series.

TABLE 1.1

Five lean principles

Lean principle	Definition
Value	Value defines what really matters to the external customer or end user. The paying customer defines the value of the product and the reason a company exists. Knowing what the customer values and understanding the business conditions for meeting this defines the day-to-day value-adding activities required for the company. Determining value is a critical stating point in thinking lean, as it determines the success of the other four principles.
Value stream	A value stream represents all the value-added and non-value-added activities required for a company's product and services to flow from concept, development, transformation of raw material, and delivery to and payment from a customer. A perfect value stream ensures that every business activity adds value to the product (customer value). Lean thinking maps the value streams to facilitate eliminating wastes and achieving flow.
Flow	Flow eliminates the non-value-adding activities in the value stream so that products or services flow continuously from concept to delivery to the customer. Continuous flow through various operations is achieved by determining the needs of the customer and the pace (takt) at which the value stream must flow to meet these needs with the least amount of delay or waiting. Just-in-time (JIT) or pull systems enable flow.
Pull	Pull is a material replenishment system initiated by consumption or by actual customer orders, wherein the upstream supplier produces something only when the downstream customer signals a need. Pull enables the value stream to produce and deliver the right materials at the right time in the right amounts with minimal inventory.
Perfection	Perfection is the ideal of eliminating all waste along the value stream to achieve continuous flow. Applying the other four lean principles enables an operation to move toward perfection through continuous improvement (kaizen).

ESSENCE OF LEAN

Because lean is, in fact, TPS, the best way to understand lean is to look to Toyota and to Taiichi Ohno: "The basis of the Toyota production system is the absolute elimination of waste. The two pillars needed to support the system are *just-in-time* and *autonomination*, or automation with a human touch."

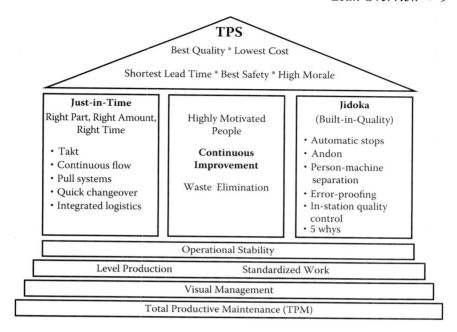

FIGURE 1.1
The Toyota Production System house.

The components of TPS are often depicted in the form of a house (Figure 1.1). The left pillar is, as Ohno had said, just-in-time, defined as making just what the customer needs, when the customer needs it, in exactly the right amount. The things needed to support that are shown, including continuous flow, pull replenishment systems, and quick product changeovers. Toyota focused on adopting Ford's original idea of continuous flow, but without the inherent waste that arises from a batch production system—mass production. It understood that to make flow waste-free it must be determined by the precise pull of the customer. So it created the concept of takt, which comes from the German word for rhythm, for the beat of a drum. Toyota realized that if it could gear the manufacturing line to produce at a rhythm in synchronization with customer demand, delivering the right parts, in the right amount, at the right time, and then the line could meet all customer needs without the waste of excess production. This is the basis for the JIT pillar.

Ohno's other pillar is generally shown as *jidoka,* or built-in quality, which is the goal of Ohno's autonomation, to employ machines that can sense any off-quality situation and stop, to avoid continuing to produce defective parts. This eliminates the waste of rework or discarding defective parts or products—another big waste of mass production—as well as ensuring that you continue to deliver quality to customer on time.

The center of the TPS house, as it is usually diagrammed, is highly motivated people, charged with continuously improving the process. This is in stark contrast with the philosophies of Henry Ford, as influenced by Frederick Taylor. Taylor had been one of the first to recognize that *planning* work and *doing* work were two separate activities. However, that novel, valuable insight led him to the inappropriate conclusion that they should be done by different groups of people, with managers and engineers doing the planning, and workers the doing. Thus, Ford's mass production system, and all that followed it, relegated workers to performing prescribed tasks without any opportunity to suggest improvements. Toyota changed all that, and encouraged, even required, line workers to continually reexamine the way their work was designed and suggest improvements. That doesn't mean the workers have a completely free hand to redefine the way their tasks are performed. Any suggestion for task redesign must be evaluated, approved, and documented so that it will become the new standard for performing that task. Standard work is a core principle of TPS, and is usually shown as part of the foundation of the TPS house.

Another foundational element of TPS is level production (*heijunka*), the leveling of production by both volume and product mix to build a schedule whereby the company produces the required amounts of each product type each day. Ohno was a strong proponent of reducing variation in production rate. By taking the monthly customer requirements and spreading them evenly over the number of days the plant was to operate, and then further dividing by the number of hours of operation and then down to the number of minutes or seconds during which each item was to be produced, Toyota could fully level out the production rate. The number of minutes or seconds available to produce each item is the takt time, the basis for the production rhythm just described. Ohno understood that a leveled schedule was necessary to maintain operational stability—another foundation of the TPS house—and keep inventory at a minimum. Otherwise, big production spikes of some products and not others would create part shortages, necessitating excess inventory.

So the essence of lean is:

- The relentless elimination of all waste
- Continuous improvement (*kaizen*)

- Manufacturing at a rate equal to true customer demand, just when the customer wants it (JIT)
- Ensuring quality by sensing defects and stopping production until the causes are found and corrected (*jidoka*)

FOURTEEN LEAN TOOLS

A key strength of TPS is that it includes not only concepts and philosophies, but also an array of effective work practices and tools to enable the concepts and philosophies to be realized on the shop floor. Fourteen key tools and work practices are described briefly here. Those requiring either slight adaptation or a significantly different point of view to meet the challenges and requirements posed by the process industries are explained in detail in subsequent chapters of this book.

Lean Tool 1: Value Stream Mapping (VSM)

VSM is a method of visually depicting your process in terms of the physical flow of material and how that creates value for your customer. The key process steps are shown, along with data related to flow, quality, lead time, and throughput capability relative to takt. Included is a diagram of how information flows and is processed to manage, control, or influence the physical material flow. A third component is a timeline, to illustrate all the things that detract from short lead times. VSMs are a key lean tool for understanding where waste is created in the process and what might be improved to reduce or eliminate it. A future state VSM is developed to show what your value stream should look like after lean improvements have been made; this provides a template for all improvement activity. Originally based on Toyota's material and information flow diagrams, the format presented by Mike Rother and John Shook in *Learning to See* has become a standard for VSMs. Enhancements to VSMs to more fully describe process industry value streams are covered in Chapter 4.

Lean Tool 2: Takt Time

Takt is the time interval at which each item, each part, subassembly, or finished assembly must be produced to exactly meet customer demand. Takt creates the pace, or rhythm, at which material must flow to meet customer needs. As covered in more detail in Chapter 4, takt is frequently expressed as a rate (pounds per minute) rather than a time (minutes per pound) in the process industries.

Lean Tool 3: Kaizen

Kaizen is the Japanese term for continuous improvement. Kaizen is a work process wherein all employees are engaged in ongoing improvement of all processes. A way of thinking and behaving, kaizen is a total philosophy that empowers the employees actually doing the work to remove waste and to design and implement more effective processes. *Kaizen events* are short, highly focused, dedicated team activities aimed at making a specific, well-defined improvement. Process–industry kaizen events are discussed in Chapter 9.

Lean Tool 4: 5S

5S is the name given to a five-step process for workplace organization, housekeeping, cleanliness, and standardized work. Its name comes from the Japanese words for the five specific steps: *seiri, seiton, seiso, seiketsu,* and *shitsuke.* The approximate English translation for these five terms is *sort, set, shine, standardize,* and *sustain.* They are described more fully in Chapter 8, on application of visual management principles in the process industries.

Lean Tool 5: Jidoka

Jidoka, or automation with a human touch, is one of the two pillars of the TPS house, and builds in quality at the source by providing equipment with intelligence to stop automatically when it senses it is producing off-quality material. Jidoka is as much a state of mind as it is the specific technology embedded in the equipment. It is a philosophy that everything must stop at the first sign of quality problems so that the problem can be corrected before production resumes, to limit the waste being produced. According to Ohno, this principle is based on a belief that a willingness to do this will ultimately lead to a line that "is strong and rarely needs to be stopped." It is supported by andons (visual indicators, usually lights) that signal the status of the line.

Lean Tool 6: Single Minute Exchange of Dies (SMED)

SMED is a process for systematically analyzing all the tasks to be performed in a product changeover, so that the changeover can be simplified and done in much less time. The methodology was developed by Shigeo Shingo, an industrial engineering consultant working with Toyota during the period of TPS development. As described in Chapter 7, SMED has been applied very successfully in the process industries not only on changeovers, but also to the annual shutdowns and overhauls typical of some of those processes.

Lean Tool 7: Poka-Yoke

Poka-yoke is a set of techniques for mistake proofing, used both to prevent defective products from being produced and to prevent production equipment from being set up incorrectly. Poka-yoke includes designing things so that they can be put together only one way, sensors to detect when things are not done correctly, and color coding to reduce the likelihood of connecting things incorrectly. Originally called *baka-yoke*, or foolproofing, the term has been changed to poka-yoke, mistake proofing, to avoid any suggestion that operators are fools. Poka-yoke is another lean technique attributed to Shingo.

Lean Tool 8: Five Whys

Five Whys is the name given to the practice of asking the question "why" five times, to get to the underlying root cause of a problem. Asking "why" a number of times is critical to this process; the number five is not. Sometimes it takes more than five queries to get to the root of the problem, sometimes less. The key is to keep asking until you understand what you must change to resolve the problem. Ohno described the Five Whys method as "the basis of Toyota's scientific approach ... by repeating why five times, the nature of the problem as well as its solution becomes clear." This practice is discussed more fully in Chapter 5, as a tool for understanding causes of waste when analyzing a value stream map.

Lean Tool 9: Standard Work

Standard work is a definition of the specific tasks to be performed by an operator, including sequence of operations and timing. In the process industries, these are often referred to as standard operating procedures (SOPs). The key concept is that there is an optimum way to perform each job, and if it can be defined and if everyone does it that way, not only will performance be optimized, but variability will be taken out of the process.

Lean Tool 10: Total Productive Maintenance (TPM)

TPM refers to a set of practices aimed at improving manufacturing performance by improving the way that equipment is operated and maintained. Highly team-based and involving all levels of the operation, it drives toward autonomous maintenance, where the majority of maintenance tasks are done by those closest to the equipment, the operators. TPM and its importance in improving asset productivity on process equipment are covered in more detail in Chapter 6.

Lean Tool 11: Cellular Manufacturing

Cellular manufacturing is the practice of dividing the full product line up into families of products requiring similar processing steps and conditions, and then dedicating specific pieces of equipment to each family. This can often lead to shorter changeovers, higher quality, reduced variability, increased throughput, and better flow. There has been a reluctance to implement cellular manufacturing in the process industries; Chapter 11 describes ways to overcome perceived barriers, an implementation methodology, and typical benefits.

Lean Tool 12: Heijunka

Heijunka is the practice of leveling the volume of material being produced over time, so that production is always to a level takt. Also called production leveling or production smoothing, heijunka increases operational stability and reduces variability in resource utilization and raw material requirements. Specific tools include heijunka boxes and heijunka boards. A tool that has been developed within the process industries to achieve the same ends is the product wheel, described in Chapter 12.

Lean Tool 13: Just-In-Time (Pull)

JIT, one of Ohno's two pillars, refers to the set of principles, tools, and techniques that enables a company to make what is needed only when it is needed and in the exact quantity needed. JIT avoids overproduction, either producing more than will be needed or producing before it is needed, thereby reducing inventories to the minimum required for smooth flow. JIT is also called *pull,* based on the principle that we will produce only what customers have pulled from the inventory shelf, which is the restocking practice commonly used in grocery supermarkets. Pull is the opposite of push production, which is driven by a forecast rather than by current customer demand. Process industry pull systems are described in Chapter 14. Solutions to problems caused by process equipment operating constraints and by high product variety are explained.

Lean Tool 14: Kanban

Kanban describes a mechanism for visually signaling what is needed—that is, what must be produced to replenish materials pulled by the customer, which may be the final customer or the next step in the process. The word *kanban* comes from the Japanese term for visible sign. Kanban has traditionally been implemented with card systems, bins, or totes marked with the quantity to be produced—that is, the lot size, and the specific product or material type. Kanban signals typically used in the process industries and methods for calculating total kanban requirements are explained in Chapter 15.

FURTHER INFORMATION

This concludes our brief description of the origins and history of lean, its key components and principles, and specific practices. For a more thorough discussion on the genesis, basic principles, and application of lean tools, I strongly encourage you to look into some of references in the bibliography in Appendix B. Key references include:

- *The Machine That Changed the World* by James P. Womack, Daniel T. Jones, and Daniel Roos (Macmillan).
- *Factory Physics* by Wallace J. Hopp and Mark L. Spearman (Irwin/ McGraw-Hill).

- *The Toyota Way* by Jeffrey Liker (McGraw-Hill).
- *Toyota Production System—Beyond Large Scale Production* by Taiichi Ohno (Productivity Press).
- *Today and Tomorrow* by Henry Ford (Productivity Press).
- And despite its somewhat demeaning title, *Lean for Dummies* by Natalie J. Sayer and Bruce Williams (Wiley Publishing) is informative, readable, and makes an excellent reference.

LEAN TODAY

Influenced by the names Toyota Production System and lean production, most early users focused these concepts and practices on the production part of the enterprise. In the ensuing years, many have followed Toyota's original practice and applied them across their internal supply chains, using the concepts to link various manufacturing processes, warehouses, distribution centers, and logistical operations in their vertically integrated, multi-echelon supply chains. They have then extended beyond their internal operations to their external supply chains, linking their operations into those of their suppliers and customers.

The more progressive lean companies now drive lean thinking throughout their entire business enterprise, taking waste out of their business processes and redefining the fundamental purpose of these processes, and integrating customer values into the entire enterprise. All functions of the business enterprise have been touched: accounting, finance, legal, human resources, and sourcing. Some companies are applying lean to their engineering function to build better plants, plants with less wasteful processes, smoother flow, and shorter lead times, and to remove waste from the project activities used to design and build plants.

In recent years, there has been a great deal of lean activity in the service industries, such as banking, education, logistics, software, and health care, a rapidly growing industry fraught with waste.

Lean is continually evolving and developing—what is called lean today is somewhat different, we hope better and advanced beyond what was described in *The Machine That Changed the World*. As lean has spread across supply chains, across entire enterprises, into all business functions and processes, and into nonmanufacturing operations, one area where application seems to be lagging is in the process industries. The literature has little to say about lean success in the manufacture of paints, toothpaste, food and beverages, paper, synthetic fibers, flooring, plastics, adhesives, or lubricants. However, some companies involved in these industries, including my former employer, were successfully transforming their operations using lean concepts even before the term *lean production* was coined. The next chapter discusses the nature of process manufacturing, how it differs from assembly manufacturing, and the unique challenges you may face when applying the lean tools you just reviewed.

SUMMARY

The body of philosophies, concepts, and tools that we know today as "lean" is based largely on work done by Toyota beginning after World War II and continuing through the 1970s and 1980s. Many of the foundational technology elements had been evolving throughout the eighteenth and nineteenth centuries, and were brought together by Henry Ford with his moving assembly line and focus on flow in 1913. Toyota emulated Ford where it could, but was forced to adapt many of Ford's concepts to fit its situation in Japan's devastated post-war economy. Toyota copied Ford's value for flow and his intense focus on elimination of all waste; added its own value for employee engagement, continuous improvement, and perfect quality; and developed work practices and tools to actualize its concepts.

As lean has continued to evolve into the twenty-first century, it is now widely viewed as comprising five key principles:

- Specify *value* in terms of the final customer.
- Identify and *map the value stream;* that is, how value is being created for the customer.
- Make the value *flow* without interruption.
- Let the customer *pull* value from the production process.
- *Pursue perfection* through continuous improvement.

Companies that have adopted and followed these principles have seen such significant improvement in their performance that lean is becoming the new manufacturing paradigm.

2

Distinguishing Characteristics
of Process Industry Manufacturing

Manufacturing processes can be categorized into two broad groups: assembly manufacturing and process industry manufacturing. Assembly manufacturing generally consists of the manufacture of individual parts and components that operators and machines weld, bolt, or otherwise fasten together into a finished product. Examples include automobiles, aircraft, motorcycles, cell phones, computers, power tools, television sets, and hair dryers. Process industries are characterized by processes that include chemical reactions, mixing, blending, extrusion, sheet forming, slitting, baking, and annealing. Process companies sell finished products in solid form packaged as rolls, spools, sheets, or tubes; or in powder, pellet, or liquid form in containers ranging from bottles and buckets to tank cars and railcars. Examples include automotive and house paints, processed foods and beverages, paper goods, plastic packaging films, fibers, carpets, glass, and ceramics. Process industry output may be sold as consumer products (food and beverages, cosmetics) but more typically as ingredients or components for other manufacturing processes.

PROCESS INDUSTRIES VERSUS ASSEMBLY OPERATIONS

Some, like Wayne Smith in *Time Out,* have characterized the difference between assembly and process industries as discrete versus continuous processing, but that is a profound oversimplification. Though many of these processes are continuous (oil refining, manufacture of bulk chemicals), many are batch chemical (house paints, industrial lubricants) or what you could consider mechanical batching (rolls of paper, tubs of fiber) and become discrete later in the process (tubes of toothpaste, rolls of carpet, quart and gallon buckets of paint, jars of mayonnaise, boxes of cereal). A more accurate characterization of the difference is that the number of different part types converges as material flows through an assembly operation, while the product variety increases as material flows through a process operation. That is, assembly manufacturing starts with a large number of raw materials and ends with a small number of finished product Stock Keeping Units (SKUs), while process operations are the opposite; few raw materials become highly differentiated as material flows through the process, ending with a large number of finished SKUs.

In *Synchronous Manufacturing*, Umble and Srikanth segment manufacturing operations into four categories:

1. **Basic producers:** Use natural resources as inputs and refine or separate them into products used as material inputs by other types of manufacturing operations
2. **Converters:** Transform materials into consumer goods or end items used by fabricators
3. **Fabricators:** Produce consumer goods or parts or components for assemblers
4. **Assemblers:** Combine parts and components to produce finished goods

Basic producers and converters are typically process industry operations, while assemblers have, of course, assembly processes. Fabricators can fall into either manufacturing category, depending on the specific type of operations performed.

Process operations and assembly manufacturing are so markedly different, with different challenges, that the Institute of Industrial Engineers has a separate Process Industry Division to focus on issues and challenges specific to that group.

These differences are profound enough that the application of lean must be approached quite differently. In fact, some have concluded that the process industries are different enough that lean doesn't really apply. In a presentation to the president and staff of one of the larger carpet manufacturers in the world a few years ago, after hearing my recommendations, the president said something like "I can see how these principles apply to Dell Computer or Harley Davidson, but not to us. We have vastly different processes than they do." I agreed that their processes were indeed different, but that the core lean principles were indeed applicable. I then proceeded to share material contained in this chapter and finished the meeting with the president's firm endorsement to proceed with the proposed recommendations.

The belief that lean only applies to assembly processes has been reinforced by the absence of any significant body of literature on application to the process industries. There are many excellent books and articles on lean (see Appendix B), but the case studies described and the successes achieved are almost entirely from assembly manufacturers.

Even Taiichi Ohno recognized that the process industries posed unique challenges. In *Toyota Production System—Beyond Large Scale Production,* he states, "To be truthful, even at Toyota, it is very difficult to get the die pressing, resin [molding], casting, and forging processes into a total production flow as streamlined as the flows in assembly or machine processing."

Ohno seems to be saying that the processes in Toyota that are similar to those in the process industries are difficult to make lean. The key word is "difficult." He doesn't say impossible.

CHARACTERISTICS THAT DISTINGUISH THE PROCESS INDUSTRIES

Wayne Smith, in his excellent book *Time Out,* addresses the "We're Different" syndrome prevalent in the process industries and concludes that the fundamental approaches of applying lean will work equally well, given an understanding of the underlying differences. Below we cover eight distinguishing characteristics unique to process industry manufacturing and highlight the uniquely different material flow patterns and dynamics found in process plants compared to parts assembly plants.

The Three Vs: Volume, Variety, and Variability

Manufacturing processes are traditionally described in terms of a volume variety continuum, with high volume, low variety production at one end, and high variety, low volume at the other. High volume production is characterized by gasoline refining, in which a company may produce hundreds of millions of gallons per year, but in only a few grades or octane levels. Custom cabinetmaking, where a shop may have only a hundred orders each year, but where each order is unique, would exemplify high variety. Assembly operations generally fall in the high volume, low variety category.

The markets served by the process industries have evolved to a point where producers must satisfy both high volumes and high variety. A synthetic fiber plant may produce 300 million pounds of fiber each year, in several hundred individual end items, or SKUs. A soft drink bottler may turn out millions of cases each year of what was once a single product, but now includes sugar-free, caffeine-free, lemon-flavored, and cherry-flavored varieties, packaged in a variety of sizes and bottle shapes.

In addition to high volume and high variety, process industries must deal with a third V, high variability. Process manufacturers may have expected relatively stable SKU demand trends in the past, but now must face the reality that the more options you give customers, the more likely they are to change preferences, so demand at the end item level can be highly variable and very unpredictable.

Capital Intensive versus Labor Intensive

Another major difference with the process industries, and one that has a significant impact on applying lean, is that these processes tend to be capital intensive whereas assembly processes tend to be labor intensive. Much of lean thinking is aimed at eliminating wasted labor and improving labor productivity. Labor productivity is an important factor in the process industries, but asset productivity is typically far more important. This is a situation where you need to expand and extend those traditional lean tools that focus on labor productivity and reducing wasted labor to shift the focus to asset productivity. Labor productivity may have been the primary catalyst leading to the development of the Toyota Production System. In Taiichi Ohno's words:

> This made the ratio between Japanese and American work forces 1-to-9. I still remember my surprise at hearing that it took nine Japanese to do the job of one American. Furthermore, the figure of one-eighth to one-ninth

was an average value. If we compared the automobile industry, one of America's most advanced industries, the ratio would have been much different. But could an American really exert ten times more physical effort? Surely, Japanese people were wasting something. If we could eliminate the waste, productivity should rise by a factor of ten. This idea marked the start of the present Toyota production system.

As a result, much of the information on a typical, traditional value stream map documents the labor required: number of operators per workstation, operator cycle time, total operator work time per piece, and so on. Collect this and you will have an understanding of the potential labor waste. However, what is missing is a breakdown of factors that detract from asset productivity, such as the number of SKUs flowing through each step, and the yield and reliability components of overall equipment effectiveness (OEE). These are critical to understanding waste and its root causes in process operations. (SKU, stock keeping unit, is used throughout the book to represent each product type, material type, or part type, that can be present at a step in the process.)

Throughput Is Limited by Equipment Rather Than by Labor

A corollary distinguishing factor in process plants is that the equipment is often the primary rate-limiting factor, rather than labor. In many assembly processes, you can eliminate bottlenecks and increase throughput by adding people. This is rarely the case on process lines. Applying more labor to a paper sheet-forming machine or a batch paint-mixing vessel will not increase throughput one iota. So the understanding of root causes of waste, and changes necessary to reduce it, must focus much more on the process equipment than how labor is applied.

As was discussed in Chapter 1, *heijunka* or leveling out production to match average takt is important for lean success. Where there is variability in demand, assembly plants can generally accommodate it by running an extra shift or working over the weekend. Process plants often run 24/7, so there are no extra shifts available to cover increases in demand. Moreover, because the equipment is usually the limitation, working overtime or bringing in temporary labor won't increase throughput. So in applying heijunka to process industries you must shift the focus to the process equipment in order to level and improve flow.

Equipment Is Large and Difficult to Relocate

You can characterize many assembly processes by their relatively small, easy-to-move machines, such as drill presses, grinders, and lathes. On the other hand, most equipment in the process industries is large and has process piping, hydraulic lines, and complex electrical wiring, making it very difficult to relocate to improve flow. Ovens used to dry extruded flakes that will become breakfast cereals can be a hundred feet long or more, and twelve to twenty feet wide. A machine to form papers used as electrical insulation in power transformers can likewise have a footprint of a thousand square feet or more, and weigh several tons. For this reason, some experts have concluded that cellular manufacturing has little application in the process industries. The opposite is true: Cellular manufacturing is one of the most powerful lean tools employed in process plants!

Processes Are Difficult to Stop and Restart

The machines typically found in assembly processes are easy to start and stop. This is not always the case in the process industries. Process equipment is often time-consuming and costly to stop and restart. For example, chemical polymerization vessels may run a two- to four-hour batch, which you can't interrupt without destroying the batch. Some continuous polymerization process lines typically run for months or years between shutdowns: stopping the flow of hot polymer causes the molten plastic to freeze in the lines. Prior to restarting, you would have to disassemble the entire process line and sandblast it clean or burn it out, at a cost of hundreds of thousands of dollars. Similarly, to stop a machine extruding plastic films causes the molten plastic to freeze in the die lip, again requiring a very extensive and costly clean-up. This tends to drive overproduction, unnecessarily large inventories, and make the implementation of a pull replenishment system difficult.

Product Changeover Issues Are Complex

In the assembly industries, product changes often involve changing tooling and then adjusting or calibrating the machine with the new tooling. The primary wastes are time and labor. In the process industries, you also make machine changes, but these are often a small component of the total waste. In addition to lost productivity, and wasted labor, there can be lost process materials, cleaning fluids and solvents, and requirements for additional testing laboratory time. For example, in the food industry, product families often have products that contain allergens, such as peanuts, and those that do not. Where volume warrants it, dedicated lines are often the answer, but in many cases, this is not economically justified. Consequently, plants must do extensive clean-ups after running the allergen-based products, with complex decontamination processes and testing to ensure a contamination-free environment afterward. In the synthetic rubber industry, process capability is sometimes poor enough to require significant production time to get properties such as viscosity back within specifications, thus generating substantial amounts of wasted material. Again, it may require significant testing laboratory time and facilities to determine when properties are finally on aim. These factors drive production schedulers toward longer campaigns.

Because process equipment is typically large and designed to be general purpose—that is, to be run under a wide range of conditions and settings—setups or changeovers tend to take much longer than typical setups in assembly plants, to allow time for the process equipment to reach the new temperature or pressure and stabilize. This also tends to encourage plants to run a large campaign on the current material prior to switching to the next. This is, of course, overproduction, and results in high finished product inventory as well as constraining flow of other products. While you can deal with some of the changeover issues in process industries using lean SMED techniques, other issues require different approaches, which are described in Chapter 7.

Finished Product Inventory versus WIP

In assembly plants, significant WIP (work in process) inventory can build up if you do not manage flow well. In the process industries, poor flow will also result in high WIP level, but in most cases, finished product inventory will also build up to a much greater level. In the process industries there is a tendency to keep things moving and, as a result, many of the ill effects of poor flow and poor scheduling (inappropriate differentiation decisions) get pushed out to the finished product warehouse.

Hidden WIP

In assembly plants, WIP often sits on the floor in bins, totes, on racks, or in carts where it is readily visible, whereas in process plants it is usually out of sight or hidden. In an architectural paint plant, the intermediate resins may be stored in 200 to 500 gallon stainless steel totes stored in an AS/RS (automatic storage/retrieval system, a high-rise storage rack system), far out of sight. Similarly, in a carpet manufacturing facility, master rolls of tufted, backed carpet, perhaps twelve feet long and six feet in diameter, waiting to be dyed, will be stored in a large rack system, well removed from the tufting, backing, and dyeing equipment. What specific material there is, where it had been processed, and where it is headed next are all difficult to determine without referring to a terminal connected to the storage management computer system. So "walking the line" of a process plant to gather flow data for a value stream map is not as effective as it is in the more visible assembly processes. This obstacle can be overcome, because there are ways to gather your data that don't require the classic hands-on tool known as "go to the *gemba*" (that is, go out to the workplace and observe) to understand the process flow.

Material Flow Patterns in Assembly and Process Plants (SKU Fan Out)

Perhaps the most dramatic difference between assembly plants and process industry plants is that the flow patterns and dynamics are actually the opposite of each other. As described in Umble and Srikanth's *Synchronous Manufacturing,* the predominant flow characteristic in an assembly plant is convergence of part types, whereas the flow in process plants is characterized by product type divergence (Figure 2.1 and Figure 2.2). Figure 2.1 represents a typical assembly plant, with material flow from bottom to top. At the start, there may be hundreds, thousands, or tens of thousands of individual screws, nuts, bolts, springs, sheets of plastic and sheet metal, and so on. As these parts are processed and assembled into subassemblies, then subsystems, then complete systems, and finally to the finished product, the number of different part types diminishes dramatically. That is, there is a significant convergence of parts into assemblies and then systems to the final product. Considering the manufacture of Toyota Camrys, for example, there may be tens of thousands of part types at the start, while

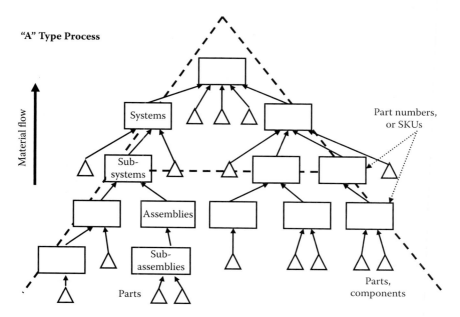

FIGURE 2.1
Schematic diagram of an "A" type process.

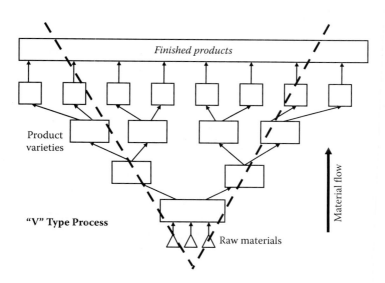

FIGURE 2.2
Schematic diagram of a "V" type process.

the final product comprises two trim lines, in several colors, for a total end item variety of perhaps a dozen or so. The flow pattern for assembly plants has been called an "A" type process, because it resembles the letter A.

Process industry operations generally follow the opposite flow pattern, sometimes called a "V" type process as shown in Figure 2.2. The process starts with few raw materials, which may be mixed, reacted, and then cast or extruded as fibers, sheets, or pellets, and then further processed to create tremendous final product variety. In the manufacture of nylon yarns for apparel or for seat belts, the process starts with few raw materials: adipic acid, diamine, TiO_2, and demineralized water. These are mixed, polymerized to create a highly viscous molten plastic, and then extruded as groups of extremely fine fibers. They can then be stretched at different ratios to build a desired tenacity level, annealed to permanently set the properties, dyed to fit the particular end use, and wound on one of a wide variety of rolls or spools. What started as four primary ingredients ends as hundreds or thousands of end item SKUs. The predominant flow pattern is one of divergence.

Examples of "V" Type Process in Process Plants

As a specific example of a "V" type process, consider a sheet goods manufacturing line (Figure 2.3). Six different grades of raw materials, in the form of small polymeric plastic pellets, are stored in input silos. A specific sheet product may be made from one pellet type or a combination of several types. Pellets are pneumatically conveyed to one of four forming machines, where a wide sheet is formed and rolled up on a "master roll." These master rolls, ten to twelve feet in width, several feet in diameter, and weighing approximately 1,700 to 3,000 pounds, are conveyed by lift truck to an automatic storage system. There can be fifty different types of master roll, with the differentiation based on pellet type, sheet width, and formed thickness.

The master rolls are retrieved from the storage system and unwound on one of four calendaring-bonding machines. The sheet is compressed into the final end-use thickness, heat-treated to lock properties in, and then wound to form what are called bonded rolls. The different calendaring pressures and bonding temperatures transform the fifty master roll types into two hundred types of bonded rolls, which are returned to the automatic storage system.

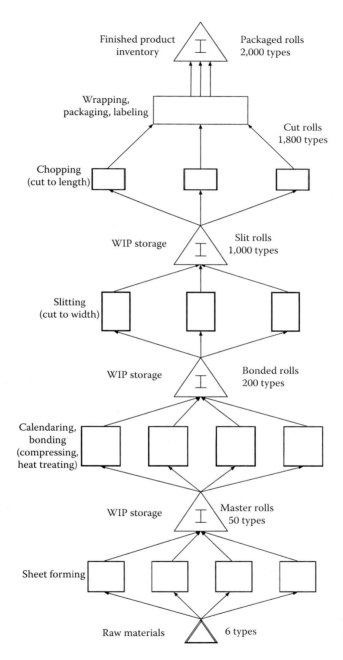

FIGURE 2.3
Example of a "V" type process sheet goods manufacture.

Sometime later, the bonded rolls are retrieved and moved by lift truck to one of three slitters, where the rolls are unwound and the sheet is passed over a rotating knife cutting apparatus. Several knives are positioned across the sheet to cut it into the desired finished product widths. The knives can be moved as required by different slitting patterns. The ten to twelve foot master roll can be cut into anywhere from two to twelve slits; slit widths can be one to eight feet. The several slit widths are automatically wound up in parallel onto cardboard cores to form slit rolls. With a high number of finished product widths available, approximately 1,000 different types of slit rolls can be created from the two hundred bonded rolls. These are returned to the automatic storage system.

The slit rolls are removed from the storage system and moved to one of three choppers, which cut the sheet to the desired length. The slit roll is unwound and rewound. At one or more times during the unwind/rewind process, a knife rises and cuts across the sheet, ending one chopped roll and beginning the next. Three or more chopped rolls can be made from each slit roll. Different chopping lengths generate 1,800 chopped roll varieties from the 1,000 slit roll types. Chopped rolls are moved directly to the finishing area immediately after they are rewound.

The finishing area consists of bagging, boxing, and labeling. Each chopped roll is automatically wrapped in a thin plastic film. The rolls are then boxed, with one or more rolls to a box depending on roll diameter. Computer-generated labels are automatically applied to the finished product boxes. A given product may be labeled more than one way, reflecting different end uses for that material, so the 1,800 chopped roll types can result in 2,000 different finished product SKUs.

So in this process flow, we have significant product differentiation at each major step: at sheet forming (6 → 50), at calendaring (50 → 200), at slitting (200 → 1,000), at chopping (1,000 → 1,800), and at finishing (1,800 → 2,000). As you can see in Figure 2.4, we have transformed six types of raw materials into 2,000 final products. This high level of divergence, of product differentiation, is typical in process plants.

Although a plant of this type may produce large quantities of product, the high variety of products results in low quantities required of some products, adding to manufacturing complexity. In the sheet goods example, total annual sales are approximately 100 million pounds. The Pareto principle applies in the extreme, with only 10 percent of the SKUs comprising

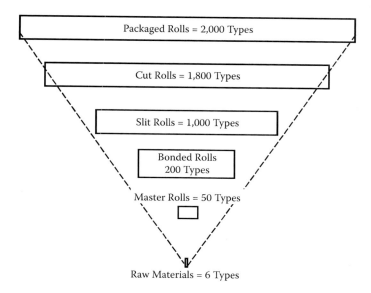

FIGURE 2.4
SKU fan out in the sheet goods example.

80 percent of the volume. So the remaining 90 percent, or about 1,800 types, comprise only 20 million pounds. Thus, some of these low volume products have only a few thousand pounds of annual sales, which requires only one to four master rolls, significantly complicating production scheduling and inventory managing processes. How to deal with this level of complexity is described in detail throughout the remainder of the book.

The same general manufacturing footprint and SKU fan out just described is found in carpet manufacture, where key process steps include tufting, dyeing, backing, cutting, and edge seaming. Significant product differentiation can occur at each step, as different carpet styles and construction are produced, then dyed to a wide variety of colors, backed with different materials, and cut to different sizes.

A similar "V" flow pattern exists in the manufacture of X-ray films. X-ray films are made in a wide array of sizes, with different photosensitivities for different end uses. A single plant would likely produce film for nondestructive X-ray analysis of welds to predict fracture likelihood, for dental X-ray, for chest X-rays, and for bone X-rays of limbs and extremities. The plant would consist of base film casting machines, coating machines to apply the photosensitive emulsions, film slitting machines to cut the wide cast rolls into more usable sizes, chopping equipment to cut the rolls into specific sheet sizes, bagging machines to place stacks of sheets into light-tight containers, and boxing and palletizing systems. A high degree of differentiation occurs at each step, as different emulsions are applied for different end uses, sheets are slit to many widths, and chopped to various sheet sizes. Packaging requirements also vary by end use. So what starts out as a small number of chemical ingredients can be transformed into hundreds or thousands of specific finished products.

Product Differentiation Points

The presence of a number of product differentiation points, where a single material can be transformed into one of several varieties, is at the core of the difference in "A" type and "V" type processes. Scheduling is critical at each differentiation point; if you make the wrong differentiation decisions, it can profoundly affect your manufacturing system performance. Your incorrect decision will produce a currently unneeded variety, which flows to finished product inventory, filling the warehouse with currently unneeded product. You have also consumed valuable production capacity, which is then unavailable to make needed products. So you end up with excessive finished product inventory, but ironically with poor customer service.

With the sheet goods example in mind, if the product is slit to currently unneeded widths, finished product inventory of those widths will increase, while shortages of other widths can occur. The same issues face other process industry manufacturers. In the manufacture of X-ray films to be used by dentists, cardiologists, and orthopedic surgeons, size requirements differ, of course, for filming teeth compared to major limbs. So the plant producing these X-ray films must make the appropriate differentiation decisions at the slitting operation, or inventory and service performance will suffer. In the fiber industry, in the elongation ("drawing") of fibers used in industrial ropes and seat belts, scheduling decisions must be made about products made at different temperatures and draw ratios, which affect end properties like tenacity and modulus. Again, you have the opportunity to make products that are inappropriate to the current customer needs.

This profound difference in flow pattern, diverging versus converging, provides both opportunities and challenges for the application of lean tools. Appropriately designed pull systems can facilitate better differentiation decisions at all divergence points, but the supermarkets in the later stages of the process must contain a managed inventory for each of the hundreds of types of semifinished product. How these and other flow issues can be dealt with is covered in the remainder of this book.

Having now been introduced to the lean principles and tools in Chapter 1 and to the contrast between assembly manufacturing and process industry manufacturing in this chapter, we can synthesize what we know so far. Table 2.1 compares and contrasts how you apply lean thinking to the two manufacturing types, but what it doesn't show is the collective impact of these differences as you move toward defining value, removing waste, and pursuing continuous improvement. The remainder of the book deals with these challenges, and with the differences in applying lean thinking to the process industries. But first, since lean application begins with driving waste out of your process, it is important that you understand the types of wastes that exist in process industries.

SUMMARY

Companies that manufacture toothpaste, house paint, salad dressings, synthetic fibers, carpeting, paper products, and bulk chemicals are all examples of what are called "process industries." Their manufacturing processes are significantly different from processes that make parts and assemble them into automobiles, refrigerators, cell phones, and the like. The type of equipment found in the various manufacturing steps of process industries is quite different from that found in parts assembly, and tends to be much larger, more expensive, and has a much higher impact on manufacturing performance and throughput. These factors make the application of many industrial engineering tools, particularly those employed in lean implementation, very challenging.

Perhaps the greatest distinction between process lines and parts assembly is the high degree of product differentiation that occurs as material moves through process operations. Assembled products are usually made from tens, hundreds, or thousands of component parts, which result in a small number of finished product types. Process operations, on the other hand, usually start with a small number of raw materials, from which hundreds or thousands of final products are made. This difference in flow

TABLE 2.1

Assembly manufacturing—process industry contrasts

Industry	Assembly manufacturing	Process industries
Examples	Automobiles Aircraft Cell phones Computers Power tools Industrial equipment Home appliances	Paints Batch chemicals Paper and plastic sheet goods Food and beverages Fibers, yarns Carpets Metals and ceramics
Process flow model	"A" processes Part variety convergence Many raw materials Little final differentiation	"V" process Material variety divergence Few raw materials High final differentiation
Prime influence and guidance	Toyota Production System	TPS with extensive application experience
Primary focus	Waste Overproduction Defects	Waste, cycle time Overproduction Yield losses
Primary economic drivers	Labor productivity Inventory reduction	Asset productivity Inventory reduction Increased throughput Reduced yield losses
Other major focus areas	N-V-A elimination Level production to takt Lot size of one	N-V-A elimination Level production to takt Lot size optimized by equipment size
Primary rate limiting factor	Labor	Equipment
Tools and techniques	Value stream mapping 5S	Supply chain mapping Value stream mapping

	Standard work Poka-yoke SMED One-piece flow Cellular manufacturing Production leveling Mixed model production Autonomation Synchronize flow to takt Pull systems	5S Standard work Poka-yoke SMED Flow determined by equipment size Cellular manufacturing Product wheels Autonomation Synchronize flow to takt Pull systems
Batch logic influenced by	Machine setup time Transportation lot size	Batch size by vessel size, roll length and width, and bale size Campaign size by changeover time and EOQ
Setup issues	Time to replace and reset tooling	Time to clean out process vessels Time to reset and stabilize temperatures Time for pressures to equilibriate Time to get properties on aim after restart
Drivers for cellular manufacturing	One-piece flow Labor utilization Flow visibility Flow management Reduced WIP Facilitate pull	Simpler changeovers Asset utilization Flow management Yield improvement Reduced WIP Facilitate pull
Cellular manufacturing implementation	Group technology Physical work cells	Group technology Virtual cells
Production leveling techniques	Control market demand Mixed model production *Heijunka*	Product wheels (batch sequence optimization and batch length optimization) *Heijunka*

pattern—divergence versus convergence—has a profound impact on flow dynamics, and on the way lean must be approached to be successful.

The remaining chapters of this book explore those differences in more detail, and show you how to adapt lean processes and tools to achieve the same or greater success as that reported by assembly manufacturers.

3

The Seven (or Eight, or Nine) Wastes in the Process Industries

Taiichi Ohno believed firmly that identifying and eliminating all waste was the key to raising productivity and becoming competitive. This idea became the genesis of the Toyota Production System.

VALUE AND WASTE

To identify waste, you must first identify *value*, because waste is anything that consumes resources (people, material, time) without creating value. The first lean principle states that value is defined by your customer, so by asking the question "what do my customers want from this process or value stream?" you are beginning the process of identifying and eliminating waste.

Ohno's experience at Toyota led him to identify seven categories of waste:

1. Waste of overproduction
2. Waste of time on hand (waiting)
3. Waste in transportation
4. Waste of processing itself
5. Waste of stock on hand (inventory)
6. Waste of movement
7. Waste of making defective parts

Ohno's categorization of wastes has endured: Most discussions of waste published today begin with Ohno's seven. The one missing element, noted by several authors (Liker, for example), is the waste of human potential and creativity, which is often referred to as the eighth waste.

Henry Ford's excellent work, *Today and Tomorrow*, gives many examples of his relentless pursuit to eliminate all waste from his processes. In fact, he has a chapter titled "Learning from Waste." Ford did not categorize waste as neatly as Ohno did, but his descriptions of waste parallel Ohno's seven.

Some have added waste of time to the list. It could be considered to be implicitly included in Ohno's list, because most of his wastes also involve the waste of time. But because time is so interconnected with all waste, it sometimes is listed as an additional type of waste. Imai does this in *Gemba Kaizen*, and this idea forms the fundamental premise of Smith's *Time Out*. Ford also considered time to be a waste, perhaps the worst waste of them all: "Time waste differs from material waste in that there can be no salvage. The easiest of all wastes, and the hardest to correct, is this waste of time." So waste of time could be considered the ninth waste, or even an overarching umbrella to cover all waste.

These wastes, as described by Ford, Ohno, and others, have been identified primarily from experience in assembly processes. These wastes also all occur in the process industries, but sometimes manifest themselves differently, and require a different way of looking at the process to identify them. They also sometimes require modified or different approaches to reduce or eliminate them.

Some of these wastes, like transportation waste, appear similar in both types of manufacturing, but have different root causes and different remedies. Others, like defects, are significantly different. In parts manufacture, defects are often caused by worn tooling, out of spec tooling, or errors in setup or adjustment. In the process industries, defects are often caused by inadequate control of key process parameters, like temperature, pressure, or viscosity, or by variability in raw material properties.

In this chapter, each waste is described in terms of the ways that it manifests itself in the process industries. Methods of dealing with the wastes are described in the remaining chapters.

WASTE OF OVERPRODUCTION

Overproduction is making more material than there are current customer orders for, which includes production by any step in a process which is currently producing more than the next step in the process requires to meet its current demand. This is every bit as much a problem in the process industries as it is in assembly industries. It has some of the same root causes in the process industries as in assembly, but also some additional ones.

As described in Chapter 2, process industry operations are characterized by significant product differentiation at most steps in the process. If scheduling decisions are made based on forecasts rather than on a demand signal based on real-time need, the wrong differentiation choice can easily be made, producing material for which there is no immediate need. In some cases, this would then go into WIP storage. When a roll of a sheet good is cast at a thickness not currently needed, it will likely go into an automated roll storage system. When a batch of a specific type of paint resin is produced before it is needed, it will be filled into a stainless steel vessel (a portable tote) and go into a storage area. For some types of products—fibers, for example—the general tendency is to keep the product moving because there is no space to store WIP, so inappropriate scheduling decisions result in excessive finished product inventory. Thus, in either case, the waste of overproduction leads to greater inventory.

Overproduction can also be the result of batch size or campaign size practices. Here the term "batch" is used to describe an amount of product made together as a unit. For example, batch chemicals are often made in amounts determined by vessel size, measured in hundreds or thousands of gallons. One vessel quantity would constitute a batch. For sheet goods, a batch might consist of one master roll, whose size is determined by the width of the forming or casting apparatus and by the capacity of the roll winding machinery. In soup production, it would be determined by the size of the mixing and cooking tanks. The term "campaign" is used to denote the number of batches of a single variety that are produced before transitioning to another variety. The mechanical equipment and the process tanks, of course, determine the ideal batch size, and the campaign length is usually dependent on the degree of difficulty in changing from one variety to another. If product changeovers are time-consuming and costly, the tendency is to run long campaigns, consisting of many batches, which of course leads to overproduction.

Another characteristic of process industry operations that drives overproduction is the unfortunate reality that many processes are not "capable," that is, not able to produce within specification on a regular basis.

Process capability is often used in process operations as a measure of the lack of product variation. When processes have low capability, it can drive overproduction: When the process is felt to be running well on the current product, there is a tendency to continue making that product as long as it runs well, rather than switch to another product that might perform poorly. Note that the process capability index is:

$$Cp = (\text{Upper Spec Limit} - \text{Lower Spec Limit}) \div 6\sigma$$

The root causes of overproduction can be seen from an appropriately constructed value stream map (see Chapter 4). Push systems (discussed in Chapter 14) will generally lead to inappropriate scheduling and incorrect differentiation decisions in process industry operations, so the push and pull arrows can highlight these situations. The information flow portion of the value stream map (VSM) should describe scheduling processes at a level where these problems are surfaced. The data box associated with each asset should also point you toward some root causes of overproduction, such as long changeover time.

WASTE OF TIME ON HAND (WAITING)

This waste traditionally includes time spent by operators waiting for the next part or lot of material to arrive, or waiting for a machine to finish processing a part so that the next part can be inserted. It also includes time spent monitoring equipment performance and part quality so that the line can be stopped to resolve any problems. The traditional lean approach to the first source of waiting waste is line balancing, reallocating the work to be done so that the processing time at each step is approximately equal, and so that the operator time is balanced to the work. The second can be reduced by what Ohno called "autonomation," designing some degree of human intelligence into the machine, so that abnormalities can be recognized and the line automatically stopped.

In the process industries, these two forms of waiting waste occur frequently, but there is often a third reason for operator waiting. In high speed, high volume processes it is not enough to recognize problems and stop the equipment; the problem must be resolved and the process restarted quickly to avoid significant lost production. Because many process plants are asset-limited, even short stoppages can result in lost production that can never be recovered.

In a synthetic fiber spinning operation, for example, fine fibers are being extruded at high speeds; several thousand meters per minute is typical. The delicate nature of the fibers makes threadline breakage a frequent occurrence. The machines have a degree of autonomation built in: A threadline break is automatically recognized and the winding process stopped. However, the molten plastic flowing through the metering pump must be kept flowing to prevent freeze-up, so material is channeled to a waste collector until the threadline can be restarted. Additionally, the production lost during the break repair can result in lost sales if the plant is operating at full capacity to meet market demand, a frequent situation. As a consequence, these fiber spinning processes tend to have more operators assigned than would be required by the normal spinning operator tasks, so that someone will always be there to repair the break immediately. The cost of the waiting waste is viewed as being minor compared to the cost of the wasted material and the potential revenue loss if the break were to go unrepaired for any length of time.

The same situation exists in the extrusion of thin polymeric films used in the electronics industry in the production of circuit boards and flexible connectors. The thin films are vulnerable to tearing during the extrusion and winding processes. As in the fiber industry, film and sheet goods processes typically involve significant operator waiting waste, so that operators are available to restart the process after holes, gaps, or tears.

The same is true with automated packaging lines that involve boxing, bagging, and palletizing equipment. If the cardboard supply is a little outside of size specifications, these lines can experience downtime. Packaging lines are sometimes the bottleneck operation in the plant, so the problem must be corrected and the line restarted to avoid a back-up that could ultimately limit production. As with synthetic fiber spinning operations and with polymeric films, this is used to justify the assignment of more operators than would be required to perform the normal tasks.

With true lean thinking, there is no excuse for this waiting waste. However, the examples cited here illustrate a mind-set and a culture that believes that this degree of operator attention is necessary, and why there is often little energy to reduce or eliminate it.

WASTE IN TRANSPORTATION

This waste includes all movement of parts and materials within a manufacturing operation. Theoretically it could all be eliminated by locating all process equipment in a line so that materials could flow from one step to the next with almost no transportation. Traditional lean approaches this by locating much of the equipment in closely coupled work cells. The process industries experience this waste to an equal or greater extent than assembly manufacturers, but the solutions are usually more difficult and much more expensive, because the equipment is so large and so difficult to relocate.

An additional transportation waste, which exists in both manufacturing models but usually much more so in the process industries, is the result of WIP being stored not on the floor between process steps, but outside of the main flow in a remote WIP storage area. In process plants this often consists of an automatic storage and retrieval system (an AS/RS, a high-rise). An AS/RS is expensive, so process plants often have a single system that is used to store raw materials and WIP from several points in the process. It is located quite far from some of the operations that feed material to it or draw material from it. This can lead to a high level of transportation waste.

The use of a large remote storage facility is an extremely wasteful practice because it results in far more than transportation waste:

- Material flow becomes quite disconnected, so people involved with the operation lose a sense of material flow.
- Inventory waste, being out of sight, tends to grow far beyond that needed for smooth, balanced flow.
- There is significant investment in the AS/RS as well as in the conveyors, lift trucks, automatic guided vehicles (AGVs), and other equipment required to move materials to and from the AS/RS.
- Manufacturing lead time, the time it takes to convert raw materials into finished products, becomes long. In fact, this is usually the largest contribution to total manufacturing lead time.

Thus, the waste being described here actually has elements of the waste of overproduction, waste in transportation, the waste of stock in hand, and the waste of time. This combination of wastes is so prevalent in the process industries and involves so many aspects of waste that it is often the greatest area of opportunity for lean application.

WASTE OF PROCESSING ITSELF

The waste of processing may include several components:

1. Building more value into the product than is required by the customer. This includes adding features to a product for which customers have no value, and packaging product to a greater extent than is required to protect the product and make it appear attractive to the customer.
2. Processing that becomes necessary to identify defects or out of spec product: inspection, testing, laboratory operations.
3. Processing that becomes necessary to correct defects or out of spec product: upgrade operations, rework, material blending.

There is relatively little of the first class of overprocessing in the process industries. Materials made to be raw materials for customer processes (such as film substrates, fibers, automotive paints, plastic pellets, sheet metal) are generally made to well-documented and well-understood customer specifications. There is, therefore, little tendency to process beyond what is required to meet the specifications. However, there are cases of overprocessing waste, for example, where powdered materials are refined more than required, and pelletized materials are screened to a higher degree than is required to remove off-size pellets. For consumer products (cosmetics, processed foods, beverages, household wrapping films), there generally is no strong tendency to build undesired features into the product nor to process beyond what is required to meet customer expectations. Of course, there are many cases where products have features that are valued by some customers but not by others, but it is often less expensive to manufacture, store, and distribute fewer varieties of more standardized products to meet a variety of customer expectations.

On the other hand, the second and third classes are likely to be more prevalent and have more negative impact in the process industries. Many of the customer-specified properties, the tenacity of a fiber bundle, the exact color of an automotive paint, the viscosity of a molten plastic, even the viscosity of ketchup, are not easy to measure. So sophisticated testing equipment and sometimes complete test laboratories are required to "process" the material to ensure that it meets specifications and to initiate product segregations and corrective action if it doesn't. These inspection, testing, and entire laboratory operations are examples of processing waste.

In some cases, when off-spec material is detected, it can be "rescued" by storing it in a rework area, and then blending it in small proportions back into the flow of acceptable product. The storing operations and blending operations are examples of the third category of the waste of processing itself.

In other cases, the off-spec material must be dissolved and filtered, or chopped up, or ground up, or shredded before it can be reused. All these shredding, grinding, and dissolving processes are, of course, examples of this family of wastes. Additionally, these processes are sometimes performed at another location, perhaps at a contractor location, so the truck loading and unloading are additional processing wastes. This also causes waste in transportation. Further, it causes waste of stock on hand in the inventory stored in the rework area. Lead time is added to the overall process by the testing, sorting, storing, and reworking processing.

This again demonstrates that many specific examples of waste show interrelated combinations of several of the nine wastes.

Some lean thinkers classify these second and third causes as part of defect waste; it seems more appropriate to me to include them as waste of processing because they often consume a considerable portion of the process capacity as well as the capacity of laboratory and support facilities. But the categorization doesn't matter as much as does the realization that they are waste, and that to be lean you must recognize them and strive to eliminate the root causes.

WASTE OF STOCK ON HAND (INVENTORY)

Theoretically any and all in-process inventory is waste. Ideally, processes should be balanced in capacity and be physically arranged in a way that material can flow immediately from one process step to the next, without any significant transportation or intermediate storage. However, there are often practical considerations that make some WIP seem appropriate. In Henry Ford's words in *Today and Tomorrow:* "It is a waste to carry so small a stock of materials that an accident will tie up production. The balance has to be found."

This and a variety of other reasons are often given to justify the need for inventory. If these reasons seem legitimate, the appropriate response is not to simply accept them, nor to arbitrarily eliminate the inventory, but to try to understand the root causes so that improvement activities can reduce the need.

There are at least five root causes of inventory waste in process plants:

Capacity Differences: Rate Synchronization

Most of the equipment found in process industry plants is large, and designed for large batches or lot sizes. Because much of this equipment is not custom-designed for its specific application, and is instead a design standardized for a variety of specific needs, the capacity, throughput, and batch cycle time can be quite different from process step to process step. Intermediate storages are required to buffer these flow differences.

For example, in a staple fiber operation, thick, ropelike bundles of fiber are spun, then stretched ("drawn"), heat set ("annealed"), cut to the length required by the end use, and packaged as bales similar in appearance to cotton bales. (Packaging synthetic fibers in the same form that cotton is packaged makes it easier for apparel producers to make polyester-cotton blends.) The fiber spinning, drawing, and cutting/baling equipment are typically designed to run at different speeds and throughputs, so there are typically more of the slower machines than the faster ones, to balance the overall throughput. The intermediate material is placed into large stainless steel tubs between each process step, and these tubs are placed into temporary storage buffers to achieve a balanced flow.

The same characteristic can be found in breakfast cereal manufacture, where the dough extrusion, flake cutting, and baking operations run at rates different from the packaging systems. For that reason, the flakes can be placed in storage silos awaiting availability of a packaging line.

Bottleneck Protection

It was explained earlier that production equipment is most often the rate-limiting factor in the process industries, not labor. In other words, the equipment is usually the bottleneck. Further, because most processing equipment tends to be large and expensive, adding an additional piece of equipment to relieve a bottleneck is generally not an economically viable option. (See Chapter 10 for more on bottlenecks.) Thus, bottleneck management becomes a critical issue. Goldratt's *Theory of Constraints* teaches that inventory buffers be used to protect a bottleneck resource from disruptions by upstream and downstream non-bottleneck resources, to maintain maximum throughput. Thus, the recommended practice for bottleneck management results in the waste of stock in hand.

Campaign Sizes

In the previous section on waste of overproduction, I discussed the tendency to run long campaigns before transitioning to another product or grade. Any overproduction will result in inventory waste, either as WIP, or as product pushed through any additional process steps to form waste in finished product inventory.

Inappropriate Product Differentiation

As has been described, a major cause of the complexity found in the process industries is the large amount of product differentiation that occurs as material moves through the process. Any step where differentiation occurs offers the opportunity for inappropriate differentiation decisions to be made, resulting in inventory. Production scheduling processes based on forecasts ("push") rather than real-time demand signals ("pull") will often result in materials being differentiated to create materials not currently needed. Thus, waste of inventory is created, which may be maintained as WIP or pushed through to the finished product stage. If there are additional differentiation steps farther downstream, it affords additional inappropriate differentiation, which can easily result in the finished product waiting even longer until genuine demand for it comes in.

Tank Heels

In the manufacture of processed foods, batch chemicals, and plastics, there is often material left in the bottom of the tank or silo after emptying it to prepare for the next batch. These "heels" result from the fact that the tank or silo design may make it impractical to pump all the contents out, so material is left in the tank from batch to batch to batch. This is generally a small portion of the total contents, but it does represent waste, and could be attacked by different tank bottom design or different pumping technology.

WASTE OF MOVEMENT

By waste of movement, Ohno was referring to the motions that people perform to execute their tasks. In the process industries, the primary waste of this type is walking. Because the equipment found in process industry plants is large, operators waste a lot of time walking from one end of a machine to the other. For example, in a papermaking process, the same operators that position rolls of paper to be fed into a bonding machine are also involved in the doffing operation (removal of the completed roll) at the other end, so movement waste is expended walking from the unwind station to the wind-up and back.

In many chemical processes the tanks and vessels are spread over an area of an acre or more. Thus, chemical operators waste a lot of time and energy walking from one part of the process to another. Many processes require a great deal of vertical space, a hundred feet or more, so operators are required to go up and down several flights of stairs many times in a day. Some chemical operators are assigned to a central control room, to monitor the process by observing computer screens driven by a distributed control system (DCS). When a task requires them to go into the field, to read field-displayed instruments or to calibrate electronically monitored instruments, movement waste can be significant.

Although movement waste, primarily walking, can be a major portion of an operator's time in many process plants, it traditionally has not gotten much attention, because it usually doesn't result in other types of waste. It rarely causes more WIP or finished product inventory; it rarely causes any overprocessing; and it doesn't directly result in the manufacture of defective product. Although it can consume considerable operator time, it rarely adds to the overall lead time required to manufacture the product, so it is often ignored.

WASTE OF MAKING DEFECTIVE PARTS

In the process industries, the analog of Ohno's waste of making defective parts is the making of material for which properties fall outside of specifications or of customer expectations.

A large portion of the defects found in assembly processes involve mechanical parts whose dimensions fall outside of stated tolerances. Materials produced by process plants can also be dimensionally out of specification: Paper and film products might lack the required thickness or width uniformity. Fine powders and plastic pellets generally have size requirements that result in defects if not met; as a result, most powder handling or pellet handling systems have devices to remove "fines" and "longs."

A large portion of the defects, or yield losses, found in process industries involve properties that are much more subtle than dimensional parameters. And most products have a variety of specifications, each of which must be met for the material to be considered first grade. House paint, for example, must meet not only color specifications but also viscosity, and even more subtle properties that can affect cracking and cratering. Fibers used in carpet manufacture must meet modulus and "denier" (a measure of fiber thickness and density) in addition to color and/or dyeability specifications. Many processed foods must meet flavor, aroma, and "thickness" or flowability (viscosity) standards as well as nutritional requirements.

In parts manufacture, defects are generally caused by worn tooling or by machinery that had been improperly set up. Once a machine begins to produce defective parts, it will continue to do so until the cause is found and corrected. In the process industry, defects can be more transient; they can come and go with the normal random variation that can occur in temperatures, pressures, line speeds, tensions, and other controlled process parameters. If the speed of a film winding machine is not precisely controlled, basis weight and other properties can vary. If the temperature of ovens used to cook processed foods is not well controlled, taste and texture can be affected. The result of this time-varying nature of defects and yield losses is that defects are more difficult to recognize, and more frequent sampling must be done. And when defects are found, diagnosis may be more difficult if the process has since come back within specification.

Detection of defects is made more difficult because the testing required often takes significant time: Test results are not always immediately available. To determine whether a batch of automotive paint is at the desired color level, a sample must be taken and used to spray a test panel, which must then be dried and graded for conformance to color specifications. To test the dyeability of fibers sold into the carpet industry, samples must be collected, wound onto skeins, dyed, dried, and then graded. The standard test procedure can take hours.

Much of the variation experienced in process industries stems from the fact that many processes begin with natural ingredients: harvested crops for processed foods, ore for metals, pigments for paints, wood pulp for papermaking, and sulfur, fluorine, and silicon for various chemical processes. These naturally occurring materials have varying degrees of purity or concentration, which requires thorough sampling, testing, and sorting or refining to prevent these variations from introducing variation into the process and adversely affecting final product properties, thus creating defect waste.

Because of all these factors, the prevention of defects and yield losses has been a major focus for most process manufacturers, but it has generally been difficult to resolve, so it remains a large area of lean opportunity.

WASTE OF HUMAN CREATIVITY

Many companies in both assembly and process industries have a culture of valuing manufacturing employees primarily for their ability to perform manual work. The fact that they have a thorough understanding of the processes they perform, can readily see many examples of waste, and have ideas for corrections and an eagerness to implement them are often overlooked.

This traces back to production expert Frederick Taylor and his realization that designing and planning work was a different activity from doing the work, and his belief that they should be done by different groups of people.

The culture that has evolved from Taylor's work is common to both types of manufacture; therefore, any lean initiative should begin by reengaging the workforce at all levels in the full improvement process. Operators, clerks, mechanics, electricians, lift truck drivers should all be included appropriately in every phase of the improvement process; kaizen events are a powerful way to begin this culture shift. The people actually doing the work have insights and experience that when appropriately engaged can result in much more accurate and complete value stream maps. Their experience and creativity can be invaluable in the development of future state VSMs. Their participation in designing visual management systems will enhance the usefulness and sustainability of these systems. The solutions they offer are frequently less complex, lower in technical sophistication and, therefore, less expensive and quicker to implement.

TIME AS A WASTE

Some lean initiatives within the process industries have focused much more on lead time reduction as the primary driver rather than focusing on waste reduction. (The term "lead time" as used here is defined as the total time that the average molecule of material spends on the manufacturing site, from the time it is received as raw material, through all the transformational processes and intermediate inventory storage areas, to the time it is loaded on a truck or railcar as finished product.) Perhaps this focus on lead time is driven by Henry Ford's view of wasted time quoted at the beginning of the chapter.

In fact, a focus on lead time reduction will drive toward the same problems and corrections as a focus on waste, because the activities that consume most of the time in any manufacturing process are all waste. Looking at it from the other direction, most of Ohno's seven wastes add time to the process. Overproduction, producing material for which there is no immediate need, of course consumes time on the manufacturing equipment. Transportation, conveying material from one process step to the next, adds time. Overprocessing, performing more steps on the product than required by the end use, consumes additional time. Stock on hand, inventory, is generally the largest component of lead time; in many process operations, more than 99 percent of the time material spends on plant is in inventory, either in raw material form, as WIP, or as finished product inventory.

Waste of making defective products consumes time in four ways in process plants. Time is consumed on the process equipment making the defective product. In the process industries, there is often time consumed by all material waiting for test results to determine whether that product is defective, which would not be required in a defect-free process. Off-spec material is then culled out and placed in temporary storage awaiting a rework process, consuming more time. Finally, the rework, upgrade, or blending activities consume still more time.

The only wastes identified by Ohno that don't add significantly to the time it takes to transform raw materials into finished products are operator waiting and operator movement. As discussed, in the process industries, equipment is usually the rate-limiting factor, not labor (there are some exceptions, such as in manual packaging operations, but these are few). Thus, anything that wastes operator time will not generally add to the overall lead time of the product.

Thus, an intense focus on lead time reduction will direct attention to most of the significant causes of waste in most process operations. So those process industry companies whose lean activity drives lead time reduction will likely arrive at many of the same opportunities identified and improvements implemented as those that focus on waste elimination. This is, for example, a major thrust of Wayne Smith's *Time Out*. And some of my Black Belt colleagues put it in a Six Sigma context: "Time is a defect."

NECESSARY VERSUS UNNECESSARY WASTE

A distinction must be made between necessary waste and unnecessary waste. Necessary waste is waste that is required to guarantee smooth continuous flow of our products to our customers, given today's process performance. Most transportation steps within the plant, whether by cart, lift truck, or conveyor, are required to complete manufacture of our materials and move them toward customers. Much of our inventory is necessary to protect flow from lead time variation, equipment reliability issues, yield losses, and unpredictable customer demand. Saying that these wastes are necessary is not to excuse them or imply that they should be ignored, but to recognize that eliminating them will be difficult. It most often will require improvement in our process performance, by stabilizing lead times, improving equipment uptime, and better demand management. Reducing or eliminating necessary waste should be a high priority, but done by examining for root cause and elimination, not by management directives to reduce the symptoms.

Unnecessary waste is waste that is present neither to maintain flow nor to ensure delivery to customers. It is often there from habit or tradition; it may have been necessary waste in the past, but when process improvements were made it was not recognized that the waste was no longer required. Sometimes it is there to give operations management a "comfort zone"; it is not required but managers don't trust process performance, perhaps due to problems in the past. Whenever process performance improves, all waste should be examined to see if any of it has become unnecessary. Unnecessary waste is generally much easier to remove than necessary waste, because all it takes is commitment and will rather than performance improvement.

SUMMARY

Lean thinking defines waste as anything that consumes resources but does not add commensurate value. Value must be defined in terms of what is important to your customers, so any operations that add time or cost to your products but don't add value as perceived by your customers are waste. Lean thinking, from Ohno's development of TPS beginning in the 1940s through today's most progressive lean companies, has focused on the elimination of all wastes. Ohno divided all waste into seven categories: Overproduction, Waiting, Transportation, Processing, Inventory, Movement, and Defective Parts. Others have added an eighth waste, the waste of human potential, to the list. Although these were categorized from the perspective of assembly operations, we find the same wastes in process operations. However, in process plants, these wastes manifest themselves differently, and often have different root causes. Root causes of waste in process operations are summarized in Table 3.1, and are contrasted with the root causes typically found in parts assembly operations.

TABLE 3.1

Root causes of waste

Waste category	Parts making and assembly	Process operations
Overproduction	Inappropriate productivity measures Long runs due to long setups Scheduling from forecasts ("push")	Large batch mentality "Economies of scale" thinking in equipment design Inappropriate productivity measures Long campaigns due to costly changeovers Long campaigns due to incapable processes Unneeded types being produced Scheduling from forecasts ("push")
Waiting	Poor workload balancing Late parts arrivals Temporary stockouts	Need for very quick response to process upsets Many tasks at start and end of a batch, but few during the batch
Transportation	Poor factory layout	Equipment scattered, not co-located Large WIP storage systems located remotely
Processing	Unnecessarily tight specifications Overspecifying requirements Making defective material	Making defective material Testing for defective material Sorting defective material Reworking defective material Preparing defective material to be recycled (e.g., chopping, dissolving)
Inventory	Overproduction To buffer against defects Unsynchronized parts flow	Overproduction Batch size differences Equipment rate differences Unsynchronized material flow Long campaigns Bottleneck protection To buffer against process upsets To buffer against demand variability
Movement	Poor process layout Inefficient workstation design Searching for tools	Process equipment large and distributed over large areas (horizontally and vertically) Central control rooms located remotely Searching for tools

TABLE 3.1 *(Continued)*

Root causes of waste

Defects	Worn tooling	Raw material inconsistencies
	Improper setups	Very sensitive processes
	Incomplete specifications	Process parameters difficult to control
	Lack of work standards	Rushing to market before products are fully developed
		Lack of work standards
Human potential	A culture of noninvolvement	A culture of noninvolvement
	Stereotypes about worker capability	Stereotypes about worker capability
	A skeptical workforce (based on both items above)	A skeptical workforce (based on both items above)

Now that you have some understanding of waste, what it looks like in process operations, and what some of the key root causes of waste are, it's time to examine your processes in detail to understand where the waste is, why it is there, and take action to drive it out. The next part of this book, "Seeing the Waste," shows you how to do just that. It begins by showing you how to develop a value stream map of your process, as a way of visually describing how you create value for your customers and how that value currently flows through your process. It will then show you how to read the VSM to understand where and why the waste is present, and how to create a vision of what a waste-free process would look like. Actions you can take to drive that waste out and move toward your future state vision are fully described in Chapters 6 through 16.

Part II

Seeing the Waste

4

Value Stream Mapping the Process Industries

In Chapter 3, I discuss the eight wastes and describe how they manifest themselves in the process industries. The question now centers on what tools are available to help us discover exactly where waste exists in our process, and on how to arrange all of our knowledge and data to help diagnose the root cause.

Value stream mapping is the answer to that question, and should be a key part of any lean improvement. For most process industry operations, the format described in works like Rother and Shook's *Learning to See* provides a good starting point. However, to describe and understand the additional complexities inherent in many process industry plants, additional features and data are often necessary. And a slightly different approach to the creation of the map is often warranted.

INTRODUCTION TO VALUE STREAM MAPPING

Value stream mapping is based on Toyota's material and information flow diagrams, and provides an effective framework for depicting the process in a way that highlights waste and the negative effect it has on overall process performance and flow. As popularized by the book *Learning to See,* it has become a standard way to describe flow, and the starting point for many lean initiatives.

A value stream map (VSM) consists of three main components:

1. **Material flow:** Shows the flow of material as it progresses from raw materials, through each major process step (machine, tank, or arrangement of vessels), to finished goods moving toward the customer. This is a high level view showing only major pieces of equipment or processing systems. All inventories along the flow are also shown.
2. **Information flow:** The flow of all major types of information that govern what is to be made and when it is to be made. This starts with orders from the customer, moves back through all significant planning and scheduling processes, and ends with schedules and control signals to the production floor.
3. **Time line:** Shows the value-add (VA) time and contrasts it with non-value-add (NVA) time. It is a line at the bottom of the VSM in the form of a square wave. This is a key indicator of waste in the process. It shows the effect of waste but not the cause; the cause should be diagnosed from the other two components of the VSM.

BENEFITS OF A VALUE STREAM MAP

There are a number of reasons why a VSM should be one of the first tasks a lean team undertakes:

- It gives the team a sense of flow, inventory, and bottlenecks. Because these are often difficult to see in a typical process industry plant, the VSM has even greater value there. It helps visualize end-to-end flow, which can be difficult to see by walking the line in a process plant.
- It provides an understanding of how value is being created for the customer.
- It brings everyone on the team to a common, shared understanding of the entire process. People generally understand their work area well, but are only vaguely aware of the details of upstream and downstream process steps.
- It highlights key areas of waste in the process.

- It depicts (usually for the first time) both the material flow and the information flow that either enables or constrains material flow, in a way that all points of interaction ("touch points") become apparent.
- It gives clues to the root causes for wastes, including mishandling of information, scheduling dysfunction, and so on.
- It forms a template for design of improvement plans. It becomes the starting point for the future state VSM.

GENERATING THE MAP

Rother and Shook, in *Learning to See,* recommend that the VSM be created by walking the line and sketching it on a pad as you go. This is often difficult in the process industries, because the physical equipment arrangement does not mirror the logical flow pattern. The equipment may be somewhat scattered over large areas of the plant floor, so it can become difficult to follow the process flow without having a clear understanding of flow beforehand. It is also sometimes true that no single individual understands process flow in detail, so the knowledge required to create the VSM necessitates getting a team together. This is not intended to diminish the value of walking the process! On the contrary, there is significant value in observing the process and interacting with the people on the floor, which shouldn't be understated. But it must be recognized that for many process industry lines, is not sufficient to "see" the process. In some cases it is better to create the skeleton of the material flow portion of the map before a thorough process walk is taken.

Our typical practice is to form a team of operators, mechanics, process engineers, and perhaps Quality Control lab operators and shipping clerks, and to convene in a conference room. It is often suggested that the VSM be drawn by hand before capturing it electronically. We have found that creating it electronically at the start is effective. We'll have a computer running a mapping/charting application like Microsoft Visio or iGrafx Professional, with the map projected onto the conference room wall, so everyone can see the additions and corrections as they are being made. Where hand-drawn charts can get sloppy and hard to read if many changes are made, the computer-generated picture is always clean. These tools make it easy to open up space if an additional step must be added, and to rearrange process boxes to maintain a smooth view of flow. Experience has taught us that because the map is easy to modify and at the same time be kept legible, participants are less reticent about correcting errors or making necessary additions. This electronic version becomes the permanent (permanent, that is, until the next change is made) record of the map and facilitates the printing of copies to be distributed to others for comment and upgrade. All of this would be a nonissue if all processes were as simple as the examples used in mapping workbooks, but the complexity found in the process industries warrants a more sophisticated approach.

DIRECTION OF FLOW

VSMs should always depict flow from left to right. Due to the complexity of some processes, there can be a tendency to locate processes on the map such that flow lines go up, down, or right to left. This usually obscures a clear view of flow.

PRODUCT FAMILIES

Selecting a subset of the full range of products made in a facility can simplify both the flow mapping and the data-gathering effort and is often recommended. However, focusing on a single product family can lead to confusion for many process industry plants. In many cases, all product families flow through the same assets, so including only a subset of the full flow can create a misleading view of how fully each asset is utilized and can hide bottlenecks. So where assets are shared across all products, the VSM should be based on the full flow, not a single family. Admittedly, this can make the VSM more complex, but this degree of complexity is necessary to get a clear view of flow. And the complexity can be dealt with so that it doesn't hide a clean view of flow.

People using a VSM to reduce waste and improve flow in their value stream should take the same view as people using a map of the United States to plan a car trip from New York to California. To create the overall plan, you don't need a street map of each city that might be on your route; you simply need a map showing the major east–west interstate highways. At some point the more detailed maps are needed, but that comes later. Interstate maps don't show traffic flow volume, but if they did it would be extremely helpful in planning areas to avoid. If, however, only part of the traffic volume were shown, it would create a misleading picture. The same is often true when depicting the flow of a single product family on the VSM rather than the entire flow.

TAKT AND CYCLE TIME

Takt and cycle time are two of the most important parameters to appear on a VSM.

Takt Time

Takt is a measure of total customer demand, expressed as a time factor. *Takt* comes from a German word meaning rhythm, or drum beat. The goal is to synchronize every part of the manufacturing operation to the rhythm of customer demand, so that customer demand can be fully met while avoiding the waste of overproduction. It is calculated by taking the time available, the total time the plant plans to be operating over some period, and dividing it by the average number of units of product that customers purchase over that time period. If, for example, a lawnmower producer has demand of 2,000 mowers per week, and runs the plant on two eight-hour shifts for five days per week, the available time is 80 hours per week, so takt is 0.04 hours, or 2.4 minutes. That is, the plant must produce a mower every 2.4 minutes to meet customer demand. If the manufacturer can synchronize all operations to the 2.4 minute takt, all customer demand can be met without any overproduction. That calculation assumes that the operation continues through lunch periods and breaks; if not, the available time must be reduced accordingly.

It is important to recognize, however, that takt may sometimes be different at each step in an operation. If there are scrap or yield losses inherent in downstream processes, the takt for the upstream processes, the time to produce each part or lot, must be reduced to accommodate those losses. If, for example, Step 4 in the lawnmower plant has a 10 percent yield loss, then Steps 1, 2, and 3 have to produce parts every 2.2 minutes to make up for the loss.

Takt may also be different from step to step if different areas run different shift schedules. It is often the case in process plants that different areas have different available times. The bottleneck operations generally run 24/7, but nonbottlenecks may run a reduced schedule. Bagging, wrapping, and packing areas, for example, may run only an eight-hour, five-day schedule. Because the time available can vary from step to step, so must the takt. In a film coating process, there may be two coaters in parallel, with significantly different processing capability. If demand for the products requiring Coater 1 is significantly less than the demand on Coater 2, Coater 1 might be run only two shifts per day, with Coater 2 running three shifts. Thus, the takt for the two coaters would differ, because of different demand levels and different available times.

Because available time does not include time the equipment is not running for lunch periods and breaks, takt will differ between areas that do and do not continue to operate during lunch and breaks.

Cycle Time

Cycle time is a companion measure of takt, a measure of the rate at which the process can produce parts, lots, or batches. Where takt is a measure of the required time within which a part or lot is needed, cycle time is a measure of the time required to produce a part or lot. Takt reflects customer demand; cycle time reflects equipment capability. So if cycle time equals takt, we have a perfectly synchronized situation, with production capability exactly matched to demand.

The cycle time measure must be increased to account for the fact that the equipment doesn't run perfectly. If a machine can, when running perfectly, produce a part every 2.0 minutes, but has a scrap rate of 10 percent, and on average loses another 10 percent of available time due to mechanical or electrical failures, its effective cycle time is 2.5 minutes. Because of downtime and scrap it can produce a good part only every 2.5 minutes, on average. Cycle time is, therefore, adjusted for several detractors: scrap and yield losses; downtime due to reliability issues; preventive maintenance (PM) time; rate reductions due to equipment or material problems; and time lost during setup or changeovers. All these factors are sometimes combined into an UPtime metric or an overall equipment effectiveness (OEE). (UPtime and OEE are discussed in more detail in Chapter 6.) So the calculation for cycle times starts with the rate at which the machine or process equipment can produce on a perfect day, multiplying that by OEE to get an effective rate of production capacity, and then taking the reciprocal to convert the rate value to a time value. So if a machine can produce parts at a rate of 30 parts per hour, and has an UPtime value of 80 percent, the effective rate of production is 24 parts per hour, giving a cycle time of 60 ÷ 24, or 2.5 minutes.

Takt Rate versus Takt Time

The fact that takt and cycle time are time-based measurements often causes confusion in process industry applications. Operators, supervisors, and process engineers are much more accustomed to dealing with rate (not time) parameters. Stating throughput as 5,000 gallons per hour means much more to them than 0.72 seconds per gallon. Stating the capacity of a film coater as 200 feet per minute is more meaningful than 0.3 seconds per foot. It may seem that specifying the capacity of a paper bonding machine as three rolls per hour is as easy to deal with as specifying 20 minutes per roll, but for many people who have been working in these processes for several years that is often not the case. Further, it is more intuitive to most people that capacity values be reduced to account for UPtime or OEE factors than it is that cycle time be increased for those factors. Even if everyone clearly understands that cycle time must be equal to or less than takt to be able to meet demand, the natural thought process seems to be that capacity should be equal to or greater than demand, so momentary mental lapses can easily cause calculation errors.

Because one of the key benefits of a VSM is to present a clearly understood picture of flow and the influencing factors, the governing principle should be to present information in the most relevant manner. For these reasons, cycle time and takt should be expressed either as time parameters or as rate parameters based on what the operations personnel are more familiar and comfortable with. The two are mathematically equivalent, the rate values being simply the reciprocal of the time parameters. Whatever the choice may be, it should be followed consistently throughout the map.

UNITS OF PRODUCTION

One of the contributors to the high degree of complexity found in the process industries is the fact that the units used to describe production volume may change from step to step in the process. In the sheet goods example described in Chapter 2, customers may receive product on slit and chopped rolls, but may order it in units of square feet or square meters. A manufacturer of protective garments may use our paperlike product as one of the laminations in its garments, and may receive it in three-feet-wide rolls, but may order and be invoiced in units of square feet.

As we move back through the production process, the units may change several times. Although sales are in terms of square feet, the sheet goods packaging area deals in units of slit, chopped rolls. The chopping machines schedule in units of slit rolls, and the slitters deal in units of master rolls.

The factors used to convert from one unit to another vary with specific product: a twelve-foot-wide master roll slit into six two-foot widths generates twice as many slit rolls as the same master roll slit into three four-foot widths. The fact that the conversion factors are not constant from product to product is another complexity often found in process industry plants.

Consider a plant making and bottling tomato ketchup. Incoming crates of tomatoes are measured in units of weight, either pounds or tons. Sugar and salt are also measured in tons, while vinegar is typically measured in gallons. So the raw material storage area has different units of measure. In the kitchen area, the material would typically be measured in gallons, and also in the pump at the first stage of the bottling line. The bottling line is rated in units of bottles, as is the label applicator. The case packer and palletizer are rated in units of cases. Customer orders are usually in terms of cases. So as we move through the process, we shift from units of tons to gallons, to bottles, to cases. The conversion factors differ based on the size of the bottle. The gallons to bottles ratio is different, of course for twenty-four-ounce bottles compared to thirty-six-ounce bottles. The number of bottles packed in each case also depends on bottle size. Thus, the units for capacity and takt will change with process step on the VSM, and the conversion factors will be based on a weighted average of the bottle sizes produced. This situation is illustrated in Chapter 10.

The question is often asked whether it is better in that type of differentiating process to base all takt and cycle time calculations on a single parameter: pounds; gallons; ounces; bottles; cases; or some other measure. There is no best answer to this question, but generally the most effective practice is to use different units for each step, with units related to the rate-determining parameter for the equipment at each step. The rate-determining parameter for the sheet forming system may be the linear speed of the roll winding apparatus: the system may be capable of producing just as many twelve-foot-wide rolls as it is of producing ten-foot-wide rolls, even though the square footage is 20 percent greater. So for the roll forming system linear feet may be the most appropriate unit; if all master rolls are the same length, rolls would also be an appropriate unit. So in that case the effective capacity could be stated in terms of master rolls per hour, which would be constant regardless of the specific product being formed.

If winding speed is the rate-limiting parameter on the bonder, master rolls would be an appropriate unit for that step. If, however, different products must be bonded at different speeds, the time to process a master roll will differ based on product type, so effective capacity stated in master rolls per hour will not be a constant. What is usually done in these cases is to state effective capacity and takt in units of master rolls, but recognize that there is variation due to product requirements. A weighted average effective capacity can be calculated based on the typical product mix.

If we know the total customer demand for a specific product in square feet, and we know that that product is formed on 10-foot-wide, 4,500-foot-long master rolls, we can convert customer takt in square feet to takt in master rolls. A different product, formed on 12-foot by 4,500-foot master rolls would have a different square-foot-to-master-roll conversion factor. We can take final customer takt in square feet and convert it product by product to the equivalent number of master rolls, and then add them up to get a total takt for the sheet forming process. This can then be compared to the weighted average cycle time, or effective capacity, to determine whether this step is a bottleneck or nearly so.

Takt and capacity can be defined in input or output quantities for any process step, whichever is most directly related to throughput capability. For example, in a slitting operation, throughput is related to number of incoming rolls to be slit, not slit rolls created by that step. It takes the same time to slit an incoming roll of a given size into two slit rolls as it does to slit it into six slit rolls. So if takt and capacity are stated in units of incoming rolls, those values will be constant regardless of the number of slits. A carton packing operation, where several units may be combined into a single carton, may have its throughput governed more by the number of completed cartons than by the number of items in each carton. The rate of a paint container filling line may be more related to the size of the tank or the number of large totes being unloaded than it is to the number of containers (pails, drums, and so on) being filled. A 1,000-gallon tote usually takes approximately the same amount of time to pump out, regardless of whether it is being pumped into 55-gallon drums or 5-gallon pails, so the takt for a filling station would be based on the number of input totes rather than the number of output drums or pails.

The governing principle should be to define takt and capacity in whatever units are most related to the throughput capability of the equipment. Of course, for any step in the process, the same units must be used for takt and for effective capacity or cycle time.

To simplify the calculations, a spreadsheet like the one shown as Figure 4.1 is often developed. It provides a framework for taking the total customer demand for each final product SKU and calculating the numbers of units that must be produced at each intermediate process step to meet this demand. Yield losses must be factored in so that the earlier steps produce not only enough to meet demand but also enough to cover the downstream losses.

WHERE TO BEGIN

Several VSM guides recommend starting at the shipping dock and working backward through the process to develop the map. For converging manufacturing processes ("A" type processes) this makes sense, is effective, and has the benefit of placing the initial focus on the customer. However, in diverging ("V" type) processes, with a high level of differentiation and, therefore, a high number of end products, it can sometimes be confusing to start at the final products. In those cases, it is easier to see flow, to describe and discuss it, by starting at the less complex front end, where few raw materials are introduced. It is often simpler, and easier to trace flow, by following the diverging paths of material as it flows through the process toward the customer. It is up to the mapping team to decide which makes more sense in the specific situation, and the team's responsibility to make sure that each step is described with the final customer in mind.

LEVEL OF DETAIL

Keep in mind that a VSM should be a high-level view of a manufacturing process. It should be done in enough detail that flow can be seen and that

SKU Number	Product Description	Average Monthly Sales (Sq FT)	Cut Roll Length (Feet)	Cut Roll Width (Feet)	Chopping Yield	Cut Rolls Required	Slitting Yield
R551100	1 × 100 Roll, Grade R55	100,000	100	1	100%	1000.0	98.0%
A533500	3 × 500 Roll, Grade A53	500,000	500	3	100%	333.3	98.0%
B475500	5 × 500 Roll, Grade B47	200,000	500	5	100%	80.0	97.5%
B476150	6 × 1500 Roll, Grade B47	50,000	1500	6	100%	5.6	98.0%

Master Roll Width (feet)	Slit Rolls Required	Bonding Yield	Bonded Rolls Required	Sheet Forming Yield	Master Roll Required	Basis Weight (oz/sq ft)	Raw Material Consumption (pounds per Master Roll)	Raw Material Required (pounds)
12	22.7	86.0%	2.2	87.0%	2.5	0.60	2025	5,114
12	37.8	84.0%	11.2	87.0%	12.9	0.75	2531	32,726
10	9.1	87.0%	5.2	87.0%	6.0	0.60	1688	10,163
12	1.9	89.0%	1.1	87.0%	1.2	0.60	2025	2,471

FIGURE 4.1

Example spreadsheet for calculating takt at each process step.

barriers to smooth flow and process bottlenecks are apparent. It should also have enough detail that the key root causes of poor flow and waste can be diagnosed. However, it should not be of such detail that the team gets swamped in so much detail that overall flow cannot be seen and understood. Going back to the interstate highway map analogy, showing local streets can make it much more difficult to follow the flow of major highways.

Where additional detail is needed, it can be shown on a more specific process map for each of the more complex process steps. These then support the higher level VSM.

PROCESS BOX

Each major step in the process will be described on the VSM by a process box. A process box may depict a single large machine or chemical process, like a bonding machine, a carpet tufting machine, or a paint mixing vessel. It could also depict a step in the process consisting of an integrated process system; for example, a carpet dyeing system consisting of dye mix tanks, heaters, pumps, and the dye application machine, or a continuous chemical polymerization system with several tanks and much process piping, could each be shown as a single process box.

It is often the case in process plants that there are several machines or vessels in parallel, each of which can perform the same process step. This can be because the overall flow requirements exceed the capacity of a single piece of equipment, or because different products have different requirements necessitating specialized equipment. Examples of differing requirements are casting widths in a sheet process, different spinning speeds in a fiber process, different reaction temperatures in a chemical process, and different final package container types for different end uses. If the parallel equipment is identical, or at least similar in all relevant capabilities, it would be shown on the VSM as one process box, with a notation that there are N units in parallel. If any of the differences in capability are significant, a process box should be shown for each as separate entities in parallel.

The number of operators normally working at this process step is shown within the process box. If people are shared between steps, fractional values may be shown. If, for example, a step has a dedicated operator and an operator shared with another process step, it could be shown as 1.5

operators for this step. In assembly processes, this can be one of the most important parameters on the VSM; in the process industries, it is usually far less important than equipment utilization.

DATA BOXES

Data boxes provide the numerical information required to understand how well material is flowing through the process, where bottlenecks or capacity constraints exist, where waste exists in the process, and to provide clues to the root causes. Data boxes quantify not only the waste, but the effects of the waste, such as inventories caused by costly setups or by overproduction.

The five types of data boxes normally shown on a VSM are described in the following sections. The typical makeup of each type is shown; for some specific situations, additional data may be relevant and should be included in the data box. There may also be cases where not all the data listed are important; in those cases, the data box should be shortened for simplicity. The intent is to show enough of the operational details to give a clear understanding of flow and its influencing factors, but not to overwhelm the map with unneeded data. It should also be emphasized that extreme accuracy is not the goal; in most cases reasonable approximations are sufficient. Where values can vary significantly, the range is often listed.

All the data listed below that are relevant to the specific process being mapped should be shown on the VSM. In some cases, the data won't be enclosed in a box, but simply shown as text along a flow line. This is commonly done with transportation data, for example.

Customer Data Box

The customer data box tells you how much material customers typically order, and how soon after an order they expect to receive it. The key parameters, as illustrated in Figure 4.2, are:

- **Total quantity per unit time:** The sum total of all products ordered by all customers, per week, per month, or in whatever time increments the data are collected. The quantity can be expressed in

Customer Data Box	
Total quantity per unit time	XX
TAKT	XX
Lead time expectations	XX

FIGURE 4.2
Customer data box.

pounds, gallons, square meters, or any other units used by the planning and scheduling processes.

- **Takt:** The time available to produce one unit of product needed to fill customer orders. It is the total time the plant is scheduled to operate during the time increment above, divided by the quantity per unit time. For example, if total customer demand is 100,000 gallons per week, and the plant runs three eight-hour shifts five days per week, takt is 120 hours per week divided by 100,000 gallons per week, or 4.32 seconds per gallon. If takt is measured on a rate basis rather than a time basis, the takt becomes 833 gallons per hour.

- **Lead time expectation:** The maximum time the customer allows from the time the order is received until material is received by the customer.

Process Step Data Box

There are a number of parameters that should be recorded for each major process step, as shown in Figure 4.3. They include:

- **Cycle time (effective capacity):** This is the time between releases of parts or batches of material from a step in the process. If a paper winding machine releases a new roll every fifteen minutes, then fifteen minutes is the cycle time. In a chemical batch process, this is equivalent to the batch reaction time plus vessel load and pump-out times. As noted earlier, cycle time can be shown as a time parameter or as a rate parameter (when shown as a rate it is often called *effective capacity*). The cycle time should be increased, or the effective capacity decreased, to account for UPtime or OEE losses. Thus, the value shown here should represent what the equipment is capable of on a typical day, not a perfect day.

Process Step Data Box	
Cycle time (capacity)	XX
TAKT	XX
Utilization	XX
Lead time	XX
Yield	XX
Reliability	XX
UPtime	XX
# SKUs	XX
Batch size	XX
EPEI	XX
C/O time	XX
C/O losses	XX
Available time	XX
Shift sched	XX

FIGURE 4.3
Process data box.

- **Takt:** The amount of time that a step in the process has to produce one unit of production so that customer demand can be fully satisfied. As discussed previously, the takt at a specific step will not be the same as final customer takt if some of the material produced at any step may be scrap or yield loss later in the process. Takt may also differ at each process step if the available time varies from step to step. From the earlier paper-winding example, the takt time might be 18 minutes per roll, or 3.3 rolls per hour as a takt rate.
- **Utilization:** This is a measure of how fully utilized a process step is, and provides a key to how close to a bottleneck or capacity constraint it may be. It is calculated as cycle time divided by takt time, or takt rate divided by effective capacity. Because many process industry production lines are asset limited, this parameter highlights that specific steps are critical to throughput. In the paper winding example, utilization would be 83 percent (15 min ÷ 18 min, or 3.3 rolls per hour ÷ 4 rolls per hour).
- **Lead time:** The time it takes for one part, one batch, one roll, one tote, or one lot to complete that process step, from the time the material enters the process step until it leaves. The lead time may include both VA time, where a transformation for which the customer has value is being performed, and NVA time (that is, wasted time).

- **Yield:** The percentage of the material entering the step that leaves with all properties in acceptable ranges for all downstream processing.
- **Reliability:** The percentage of time that the equipment is not down because of equipment failure.
- **UPtime:** A metric that encompasses all forms of time lost: reliability downtime, PMs, yield losses, setup or campaign changeover time, rate reductions, and so on. It is similar to OEE (overall equipment effectiveness). A more complete discussion of UPtime and OEE is included in Chapter 6.
- **Number of SKUs:** The number of specific product types leaving this process area. The number should account for any and all differentiating features. Where the data boxes show many more SKUs leaving a process step than entering it, that signifies a highly differentiating step. If a process step has equipment in parallel, and specific products require specific machines or vessels, the SKU count for each should reflect the portion of the total SKU number processed on that machine or vessel.
- **Batch size:** The amount of material produced as a single lot or batch. Examples of a single batch are a single roll of film, paper, or carpet; a quantity of dough made as a unit, although it may exit this step as a number of loaves; a mixing vessel's quantity of paint; a reactor's quantity of bulk chemicals. This should not be confused with campaign size, where several batches of one SKU may be run before changing to produce another SKU.
- **EPEI:** A lean term for "every part, every interval." It is the time span over which all, or almost all, product types are made. There are often cases where low volume products are not made every interval; in those cases, EPEI would be the time span over which all the high volume products are made. If product wheels (Chapter 12) are being used, EPEI is equal to wheel time.
- **C/O time:** Changeover time, the time to change from one product type to another, including the time to get to full rate on the new product and get all properties within quality specifications. Sometimes referred to as "good product A to good product B time"; synonym: setup time.
- **C/O losses:** The amount of material lost in a product change. This is usually the amount of off-spec product made getting properties back within spec after the mechanical changeover is complete. In the

process industries, material losses often have more of an impact on financial performance than the time lost does. Showing this on the VSM helps clarify the reasons for long campaigns.

- **Available time:** The total time this process step is scheduled to run. If, for example, a process step is run for two eight-hour shifts for five days per week, the available time would be eighty hours per week. If the equipment is not run during lunch periods and breaks, this might reduce the available time to seventy hours per week. In the process industries, different process steps may have different shift schedules, so available time could vary from step to step.
- **Shift sched:** The numbers of hours per shift, shifts per day, days per week, or the total number of hours per day and number of days per week. For example, $8 \times 2 \times 5$, $12 \times 2 \times 7$, or 24×7.

There may be cases where some of these data are not relevant, in which case it wouldn't need to be shown. And in other cases, to capture complete understanding of flow, additional items may be listed.

Inventory Data Box

The inventory data box (Figure 4.4) is used for all inventories: raw materials, WIP, and finished product inventory. Total inventory is shown, without distinguishing safety stock and buffer stock from cycle stock. (These components of inventory are defined in Chapter 15, and methods for determining those stock requirements are explained.)

In addition to showing inventory level in units of production volume (such as pounds, gallons, square feet, cubic feet), it should be shown in days of supply. This is equivalent to the average lead time through the inventory, and also shows how long it takes to consume that volume of inventory. Alternatively, inventory turns could be shown, but that generally doesn't convey the effect of inventory on flow as clearly.

Inventory Data Box	
Average inventory	XX
Days of supply	XX
# SKUs	XX

FIGURE 4.4
Inventory data box.

- **Average inventory:** The average total amount of inventory of all SKUs stored at that position in the process flow. In the process industries, an inventory location may be shared by WIP from several points in the process, so what should be shown here is only the inventory at that stage of production. This should be in units of material volume: pounds, gallons, square feet, rolls, and so on.
- **Days of supply:** This is the inventory volume converted to a number of days. This can be calculated using a variation of Little's law: Days of supply = WIP ÷ Throughput.
- **Number of SKUs:** The total number of SKUs or product varieties normally stored at that stage of the process flow. Because the primary flow characteristic of many of these processes is divergence, based on a high level of product differentiation, this value is important to the understanding of flow dynamics.

Transportation Data Box

For all transportation steps, including deliveries from suppliers of the most significant raw materials, and deliveries to warehouses, distribution centers, and customers, the following data (Figure 4.5) should be shown on the VSM:

- **Delivery frequency:** How often shipments are received; how often shipments to warehouses and customers are made.
- **Lot size:** The average quantity shipped or received. This should be listed in the same units used in order and production data (pounds, gallons, square meters, bales) not as a number of truckloads or railcars.
- **Transport time:** The average time from shipment to receipt.

Transport Data Box	
Delivery frequency	XX
Lot size	XX
Transport time	XX

FIGURE 4.5
Transportation data box.

Supplier Data Box	
Order lead time	XX
# SKUs	XX

FIGURE 4.6
Supplier data box.

Supplier Data Box

Not all raw material suppliers should be shown on the VSM. Only those who supply a significant quantity of the total raw material volume should normally be shown. If, however, there is a low volume raw material with lead times or delivery performance that can likely affect material flow through our manufacturing process, it should be shown to highlight that potential constraint. A typical supplier data box is shown in Figure 4.6.

- **Order lead time:** The time from placing the replenishment order until the supplier ships it. The remaining component of total lead time will be listed as transport time on the transportation step. If our supplier is using a make-to-stock (MTS) strategy, the order lead time may be short—a day or two. If, however, the supplier employs a make-to-order (MTO) strategy, not uncommon in the process industries, order lead time may be several weeks.
- **Number of SKUs:** The number of material or part types we normally receive from that supplier is useful information when analyzing our raw material replenishment strategy.

Other relevant information, such as lot size normally ordered, will be listed with the transport data. The icons most commonly used for the material flow portion of the VSM are shown in Figure 4.7. Figure 4.8 shows a portion of the material flow on a VSM with the appropriate data boxes.

INFORMATION FLOW

The top half of a VSM depicts the flow of all information that schedules, manages, and controls the physical material flow. One of the major strengths of the value stream mapping technique is this view of how material flow is enabled, or constrained, by how the information is being

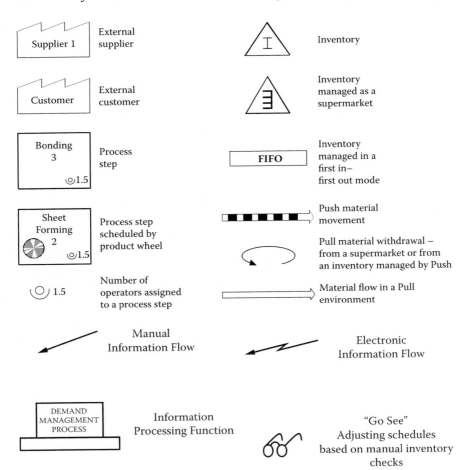

FIGURE 4.7
VSM icons.

processed. Toyota realized this when developing the material and information flow diagram concept, on which the now commonly accepted VSM is based.

We have found that in many cases, material flow is limited not by physical bottlenecks inherent in the process equipment, nor in flow problems related to equipment performance, but by mismanagement of demand data, customer orders, and production schedules. When this happens, it creates what are sometimes referred to as capacity constraints—see Chapter 10. So showing material flow and information flow on the same map highlights the points of interaction and can focus attention on those that disrupt flow and create

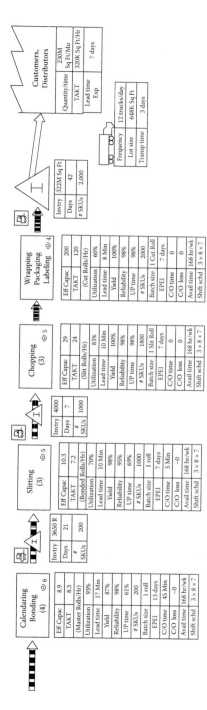

FIGURE 4.8
VSM material flow with data boxes.

Customers, Distributors

Quantity/time	230M Sq Ft/Mo
TAKT	320K Sq Ft/Hr
Lead time	7 days
	Exp

Frequency	12 trucks/day
Lot size	648K Sq Ft
Transp time	3 days

Invtry	322M Sq Ft
Days	42
# SKUs	2,000

Wrapping Packaging Labeling ☺ 4

Eff Capac	200
TAKT	120
(Cut Rolls/Hr)	
Utilization	60%
Lead time	8 Min
Yield	100%
Reliability	98%
UP time	98%
# SKUs	2000
Batch size	1 Cut Roll
EPEI	7 days
C/O time	0
C/O loss	0
Avail time	168 hr/wk
Shift schd	3 × 8 × 7

Chopping (3) ☺ 5

Eff Capac	29
TAKT	24
(Slit Rolls/Hr)	
Utilization	83%
Lead time	10 Min
Yield	100%
Reliability	98%
UP time	98%
# SKUs	1800
Batch size	1 Slit Roll
EPEI	7 days
C/O time	0
C/O loss	0
Avail time	168 hr/wk
Shift schd	3 × 8 × 7

Invtry	4000
Days	7
# SKUs	1000

Slitting (3) ☺ 5

Eff Capac	10.3
TAKT	7.2
(Bonded Rolls/Hr)	
Utilization	70%
Lead time	10 Min
Yield	98%
Reliability	95%
UP time	69%
# SKUs	1000
Batch size	1 roll
EPEI	7 days
C/O time	5 Min
C/O loss	~0
Avail time	168 hr/wk
Shift schd	3 × 8 × 7

Invtry	3650 R
Days	21
# SKUs	200

Calendaring Bonding (4) ☺ 6

Eff Capac	8.9
TAKT	8.3
(Master Rolls/Hr)	
Utilization	93%
Lead time	17 Min
Yield	87%
Reliability	98%
UP time	61%
# SKUs	200
Batch size	1 roll
EPEI	13 days
C/O time	45 Min
C/O loss	~0
Avail time	168 hr/wk
Shift schd	3 × 8 × 7

waste, such as breakdowns in communication, where people with a role in planning or scheduling processes do not receive needed information.

Creating the information flow portion of the VSM should start by showing the flow of all incoming customer data, including actual orders and schedules of future demand. The flow of this incoming information is then traced through all the transactional processes involved in creating daily and longer-term production schedules. Any batching of incoming information or delays in processing should be noted. In electronic ordering processes (electronic data interchange, or EDI) information is often processed in real time, but sometimes the systems involved update only once per twenty-four hours. In more manual processes, orders are sometimes accumulated in weekly batches before processing. Thus, the information processing itself can consume valuable time. Any such delays consume part of the available customer lead time and, therefore, allow less time for order fulfillment by the manufacturing process. In make-to-stock situations this may not be significant, but in make-to-order (MTO) or finish-to-order (FTO) processes it can be critical and may even eliminate the potential for MTO or FTO (see Chapter 13).

The information flow portion of the VSM should show each significant information processing step as a box, indicating what group performs that step, computer applications such as MRP (Material Requirements Planning) that may be involved, and whether it is a real-time, daily, or weekly process. These information boxes are connected by arrows: The usual standard is that zigzag arrows depict electronic information flow and straight arrows depict flow by paper, telephone, or fax. Each arrow should be labeled with a brief (two- to four-word) description of the content of the information flow. Figure 4.9 shows a portion of the information flow shown on a VSM. Figure 4.10 shows it coupled with the process steps depicted in the material flow.

The flow depicted by these arrows should start with customer input, move through the various information handling processes, and terminate with arrows to the various material flow processes that are scheduled. Information arrows should also flow back to suppliers, reflecting how raw material replenishment orders get created and communicated.

THE TIMELINE

The timeline appears as a square wave at the bottom of a VSM, and is intended to contrast NVA time and VA time. In many assembly processes,

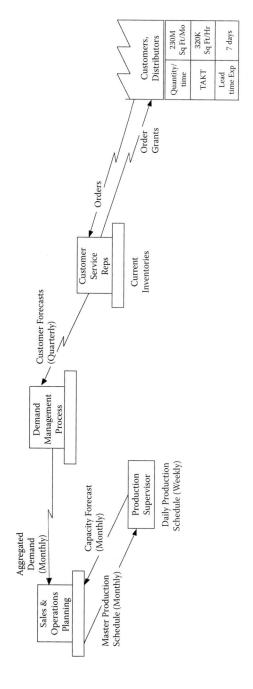

FIGURE 4.9
Information flow.

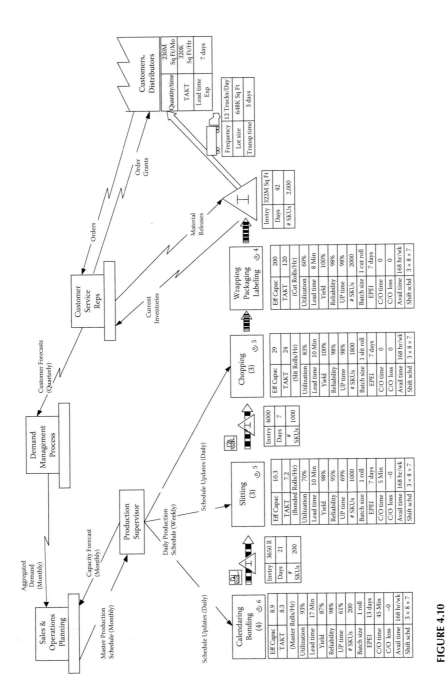

FIGURE 4.10
Information flow coupled with material flow.

the NVA time occupies 90 percent of the total time material is on the plant. In the process industries, this ratio often hits 98 to 99 percent. The timeline highlights the areas where waste adds to overall plant lead time. As we saw in Chapter 3, most of the eight forms of waste add time so the timeline is an indicator of the presence of most waste.

The normal convention is that the NVA time is the positive portion—the top—of the square wave, while the VA time is the negative (or lower) part of the wave, although this is not completely standardized across the lean community. The timeline is occasionally shown with the NVA time as the lower portion of the wave. A portion of a VSM timeline is shown in Figure 4.11.

It is important to keep in mind that the reason for the timeline is to indicate major areas of time and, therefore, waste. Extreme accuracy is not needed, so don't spend a lot of time trying to refine the data for the timeline. If the material flow portion of the map has been properly depicted, all the data necessary to draw the timeline are already included in the data boxes.

When parallel pieces of equipment or process systems have different lead times, VA times, or NVA times, the timeline should be based on a weighted average of the VA time and a weighted average of the NVA time, with the weighting factors based on the proportion of total flow volume through each piece of equipment.

AN EXAMPLE VSM

A completed current state VSM of the sheet goods process shown in Figure 2.3 would look like Figure 4.12.

ADDITIONAL VSM BEST PRACTICES

There are two other best practices to follow that maximize the usefulness of your VSM.

Parallel Equipment

As discussed, when a step in the process has two or more pieces of equipment (machines, tanks, reactors) or systems (dyeing systems, extrusion

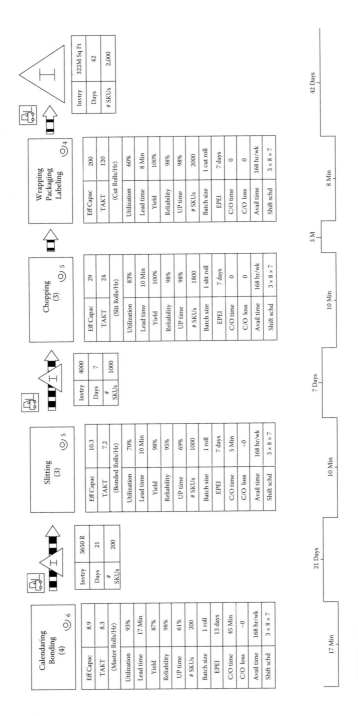

FIGURE 4.11
The VSM timeline.

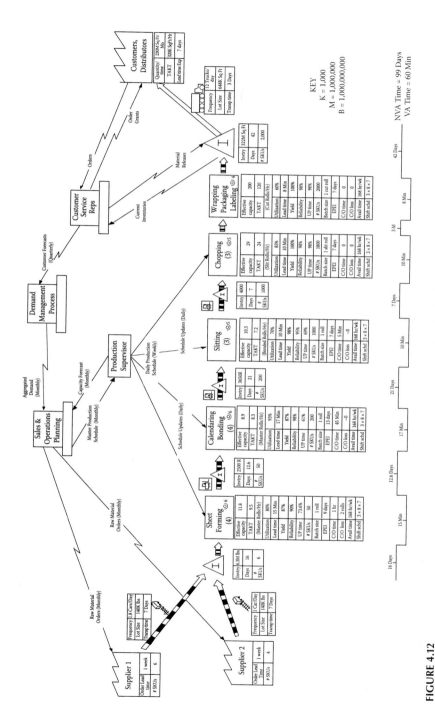

FIGURE 4.12

A completed VSM of the sheet forming process.

systems) in parallel, the recommended practice is to show a single process box and a single data box for that step, with an indication of the number of parallel units.

If the parallel equipment or processing systems have different characteristics that influence material routing or processing in any way, the parallel units should be shown as separate process boxes, with separate data boxes. In our sheet goods process, each of the bonders may be capable of bonding at different temperatures in a way that requires some material to flow through a specific bonder based on the properties required by the end user. In the sheet chopping area, some choppers may be capable of processing only the narrower range of rolls coming from the slitting operation, while others are designed to handle the wider sheets.

In some situations, parallel equipment is designed for different throughput capacity, where the higher speed equipment is reserved for high volume products and the slower equipment for the products with lower market demand.

Suppose that two of the three choppers in our sheet goods process are capable of handling any of the possible roll widths, while one chopper is designed to handle only narrow rolls, say one foot to three feet in width. The VSM for that step would look like Figure 4.13.

Notice that we have divided the effective capacity based on the individual machine capacities. Additionally, the takt has been allocated to the choppers based on demand for each width range. This results in the utilization of the narrow roll chopper to increase to 89 percent, while the universal choppers dropped to 80 percent. If this became a problem, some of the narrow rolls could be chopped on the universal choppers.

In some cases, the different pieces of parallel equipment may have different reliability history, different yields, or even be staffed on different schedules (specialized equipment needed for only a small portion of the overall flow may need to be staffed only eight hours per day, for example). The data boxes should reflect all these differences.

Logical Flow versus Geographic Arrangement

For a VSM to provide the appropriate insight into material flow, and into any detractors to smooth, synchronized flow, it is vital that flow be shown on a logical rather than on a geographically correct basis.

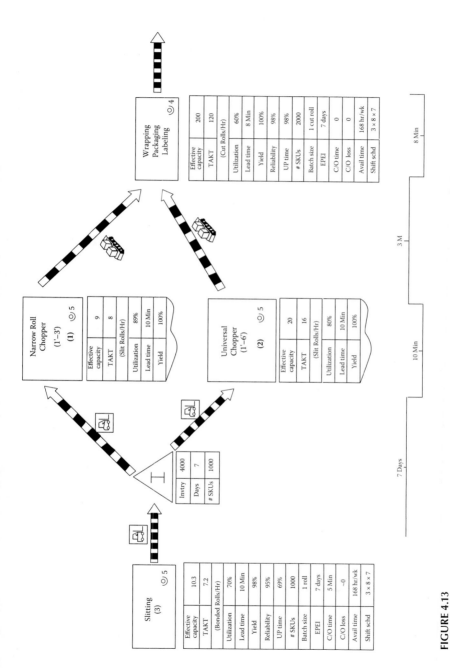

FIGURE 4.13

Parallel equipment with different capability.

There is sometimes a tendency to arrange the process boxes on the VSM in approximately the same relationship as their physical arrangement of the plant floor. This can have a useful purpose in highlighting the transportation waste, and its influence on transportation lot size and, therefore, production lot size. It may also have a benefit in understanding opportunities to relocate equipment to reduce transportation waste, although in the process industries equipment relocation is usually economically impractical due to equipment size and connections.

However, any benefit gained by depicting a geographically accurate layout is minor compared to the loss of a clear view of logical flow, which is vital to our understanding of the transformations being made to material to meet customer requirements, and the wasteful things being done along the way.

As an example, in our sheet goods process (refer to Figure 4.12) there are three WIP locations shown: between forming and bonding; between bonding and slitting; and between slitting and chopping. These are actually located in a single AS/RS (automatic storage/retrieval system), a high rise rack system using automatically guided cranes to transport a roll to the designated slot. A VSM based on the geographic arrangement would look like Figure 4.14.

Although geographically accurate, this view provides no sense of material flow or the sequence in which specific operations are performed.

The fundamental principle is that boxes on the map should be arranged so that true flow can be seen. If storage areas and testing labs must be shown several times on the map to accomplish this it should be done.

There is often a need to see flow on a geographic basis, especially where some equipment relocation may be possible, and that should then be depicted on a separate map.

A properly drawn VSM is an extremely important component of any lean activity. It provides a detailed understanding of the current state, in a way that clarifies flow and detractors to smooth flow. It accurately depicts the major effects of waste and wasteful processes, and provides insight into root causes of waste. It is the starting point for creation of a vision of what the future state should look like: the future state VSM. It provides a template and design information for application of lean improvements like cellular manufacturing, production leveling, and pull replenishment systems. It provides a context for prioritizing all the improvement initiatives arising from analysis of the VSM. It should be the first work activity

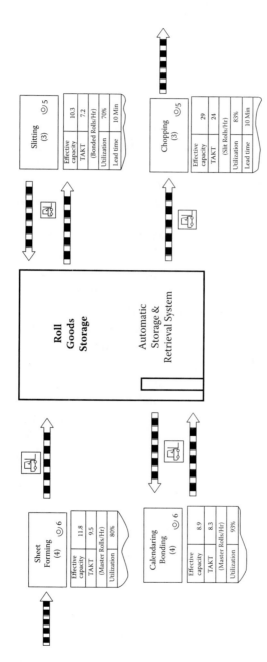

FIGURE 4.14

Geographic representation on a VSM.

in any lean transformation, begun as soon as the organizational work and team formation are completed.

SUMMARY

Value stream mapping is a powerful tool for understanding how you create value for your customers, how waste is being created along the way, and what process steps and managing processes are inhibiting smooth flow. It should be developed early in the lean implementation, as soon as a team has been formed and before any improvement tasks are begun.

To completely describe process industry manufacturing operations, a VSM must include more than a VSM for an assembly process typically would. Because equipment is very expensive and often the rate-limiting factor in process lines, equipment should get more focus on the VSM. OEE and its components, as well as changeover losses and changeover time, also have greater importance. Because process industry operations tend to be highly divergent, with significant product differentiation occurring at one or at several steps, the number of product types or SKUs leaving each step is critical to understanding why the system is being managed the way it is and why it is performing the way it is.

Because materials are transformed from a liquid state to a solid state, and then to a solid in a different form, in many processes, your choice of the units of volume (gallons, pounds, square meters, or rolls, for example) to use at each process step is important. Whether to describe takt and capacity in time units or rate units is another important consideration.

All these factors must be considered—and the best choices made—for your VSM to be an effective guide for your lean journey.

5

Reading and Analyzing the Current State Value Stream Map

An accurate VSM can provide three useful and valuable functions:

1. Bringing a higher level of clarity on how the process is currently performing and why
2. Highlighting areas where improvement should be made
3. Providing a template for documenting the ideal future state

If the VSM did only the first of these, it would be well worth the time and effort that went into its creation. However, it can do so much more that it is a mistake to stop at that point.

Many references recommend that after completing the current state VSM, the next step is to create a future state VSM. They suggest that decisions on where within the process to work toward continuous flow and which step(s) to schedule should be made, and a pull strategy can then be developed. Although that will sometimes work, there are frequently enough process issues, variabilities, and instabilities, that going directly to pull, or to any desired future state can be difficult. Achieving the future state can be much more feasible if key performance detractors are identified and resolved, or integrated into the future state plan.

ANALYZING THE CURRENT STATE MAP

So before jumping to creation of a future state VSM, you should analyze the current state map to see what it can tell you about your process, its current performance, and opportunities to reduce waste. The more obvious future state goals are much easier to reach if the process has first been stabilized and improved. Thus, the move to a defined future state should be more than going to pull replenishment and continuous flow; it should include correction of all significant performance detractors present in the current state, which can be found through a careful analysis of the current state VSM.

Several "lenses" can be used to read the current state VSM. In each case you are looking for similar things, but using a different frame of reference or set of questions to guide the analysis. Each of these will reveal many of the same problems or causes, but it is useful to examine the current situation from several different points of view to build a more complete understanding of current detractors and potential causes.

Voice of the Customer

The first view is that of the customer. In Six Sigma terms, "what is the voice of the customer telling you?" Look at customer facing metrics, including:

- Delivery performance
- Service levels
- Number of defects found by the customer
- Invoice accuracy

Waste

Look for where the waste is in the process. From Ohno's list of seven wastes:

- **Inventory:** The triangle icons on the VSM identify where inventory is being held. Knowing where it is being kept should raise questions on why, on what the root cause requiring the inventory is.
- **Waiting:** Although this is generally not a big factor in the process industry, it may be in specific cases, so it should be examined. The symbol showing the number of equivalent people assigned to each step, along with an understanding of the value-adding tasks for that step, should provide some insight into waiting time for that step.
- **Transportation:** A good VSM should clearly identify all transportation required to enable material to flow through the process, using

the lift truck, conveyor, and other transportation icons. Although process industry equipment is not always practical to relocate to reduce transportation, questions about what could be done to reduce it should be raised.

- **Overprocessing:** This includes "rescue" activities, tasks that are done to take out of spec material and perform additional processing to make it acceptable. These often show on the VSM as "upgrade" or "rework" steps.
- **Production:** Making more than is currently needed will be shown as WIP or finished product inventory.
- **Defects:** The yield field in process step data boxes will identify the magnitude of defects. The causes usually require more analysis.
- **Movement:** People movement is generally not obvious from a VSM. If it is suspected that this is a significant waste, draw a point-to-point chart, which is a plan view diagram of the work area, with lines drawn to indicate all the movements operators make as they walk around the work area performing various tasks during the day. When completed, all the many crossing lines usually resemble a plate of spaghetti, so this chart is frequently called a spaghetti diagram. If the diagram has that appearance, it is an indication that there is significant movement waste.

Non-Value-Adding Activities

Look for non-value-adding (NVA) tasks and activities. This is synonymous with looking for waste, but some people can understand and recognize NVA tasks more readily than waste.

Flow and Bottlenecks

Look for flow and continuity issues: "Why don't I have continuous flow?" and "Why don't I always make to takt?" In addition, does the process have one or more bottlenecks? Potential bottlenecks? Any step with a utilization of 100 percent or more is a step that is not able to make takt, so you won't be able to completely satisfy your customers. Any step approaching 100 percent utilization might not be able to make takt on a regular basis.

Variability

Look for variability in the process:

- Variability in supply lead time
- Variability in manufacturing lead time

Other Opportunities

Look for opportunities to apply:

- TPM (see Chapter 6)
- SMED (see Chapter 7)
- Cells (see Chapter 11)
- Product wheels (see Chapter 12)
- Pull (see Chapter 14)

Table 5.1 summarizes a list of questions to pose as you go through this process.

TABLE 5.1

Thought-provoking questions

1 Where on the VSM timeline are the long NVA times?
What is the cause of the long time?
How can it be reduced or eliminated?
2 Where are the push arrows on the VSM?
What would it take to convert these to pull?
Are scheduling decisions being made from forecasts or from real-time data?
3 Where in the process are the highly differentiating steps?
Are differentiation decisions made on current need or on a forecast?
Does this lead to inappropriate differentiation decisions?
How much inventory does this cause?
4 Which process steps have long campaign cycles (high EPEI)?
What can be done to shorten campaigns?
Which process steps have long or expensive changeovers?
How is product sequence being optimized in these steps?
Has SMED been applied to steps with long setups? Again? Again?
5 Where is inventory in the process?
Is it due to adjacent process steps having different rates? Can rates be better synchronized?
Is the inventory to protect a bottleneck? Can the bottleneck be relieved?
Is inventory being carried to accommodate long campaigns? See Question 4.

TABLE 5.1 *(Continued)*

Thought-provoking questions

6 Which process steps have high yield loss?
   Is excessive inventory being created to compensate?
   What is the root cause of the yield loss?
   Where is the defect discovered?
   Could the cause of the defect be eliminated?
   Could the feedback loop be shortened?

7 Which process steps require testing for adherence to specification?
   Is this testing being done by inline instrumentation or by a test lab?
   Is the lab in a remote location or located within the process flow?
   How long does it take to get test results? To get results to the operation?
   How can lab response time be accelerated?
   Is the technology available to test inline?

8 Are there upgrade or rework areas?
   Could the cost of operating the rework area be rechanneled into development work to
      eliminate the need for rework?
   Is out-of-spec material blended back into the process flow with first-grade material?
   Does it have to be prepared (cut, shredded, dissolved, ground up) prior to reuse?
   How do the costs of testing, sorting, dissolving, grinding, and recycling compare to the
      cost of solving the root cause of the quality issue?

9 Where do you see the material handling icons (fork trucks, lift trucks, etc.) on the VSM?
   How could this transportation be eliminated?
   Is there the possibility to relocate equipment for smoother flow?
   Are storage areas out of the main product flow?
   Could these storage locations be relocated closer to the main flow?
   Could the need for the storage be eliminated?
   Could the need for the storage be minimized to allow relocation within the process flow?

LEARNING FROM MATERIAL FLOW

The material flow portion of a VSM can reveal where waste exists in the process, and give clues as to why.

Figure 5.1 shows the raw material inventory for the sheet forming process, and indicates that the current quantity is 6.3 million pounds, about sixteen days of supply. That appears to be high, realizing that materials are received daily. Given all the variables likely to be present in this situation, a raw material inventory of less than four days should be sufficient. (See

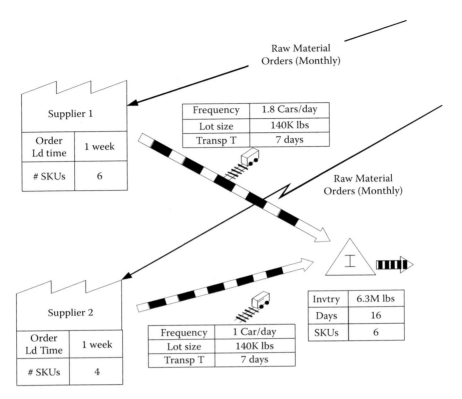

FIGURE 5.1
Raw material inventory.

Appendix A for details on how to estimate that.) So there is not only waste, but an excessive amount of it, four times the amount that might be needed with better operating practices. To understand why such an excess is there, a look at the information flow portion of the map shows that materials are ordered monthly, based on a forecast rather than on current quantities and needs. An ordering system based on actual raw material consumption rather than on a forecast, and done more frequently, should enable this inventory to be dropped to a fraction of the current value.

Figure 5.2 shows the portion of the VSM with the slitters, choppers, and the WIP in between. It shows that the WIP is currently at 4,000-slit rolls, about seven days worth. The reason is likely that both the slitters and the choppers run a seven-day product cycle (shown as EPEI in the data boxes); if their schedules are not coordinated, almost that much inventory would be required between them. Finding practical ways to reduce the slitter

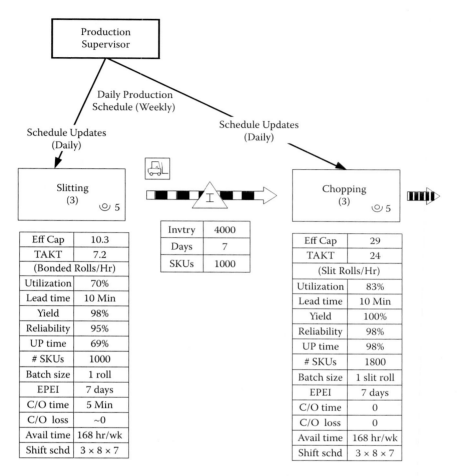

FIGURE 5.2
WIP inventory.

campaign size, including SMED activities, should enable a reduction in this WIP. Better coordination between slitting and chopping might allow it to be almost eliminated.

The VSM indicates that there are no bottlenecks in the process: All process steps have utilization less than 100 percent (after taking into account all yield and reliability factors). However, the bonders could be a potential bottleneck. Figure 5.3 shows that the four bonders have a collective utilization of 93 percent. Although this indicates that there is still time available (7 percent), there may be periods of time when the bonders cannot meet takt. The yield (87 percent) and reliability (98 percent) numbers tend

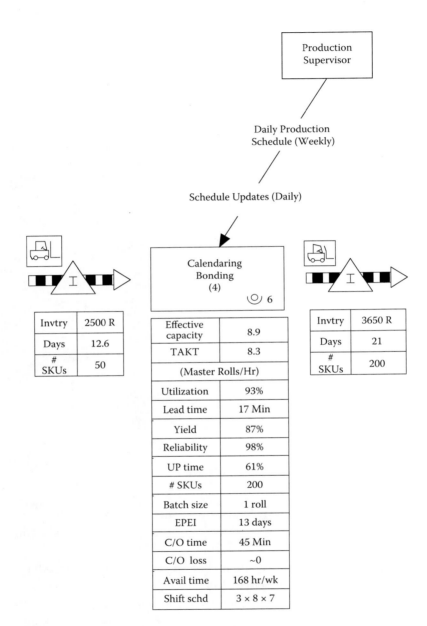

FIGURE 5.3
Potential bottleneck.

not to be constant with time, but to vary with time and with the product being run. The 87 percent yield number is an average, and could be 97 percent for some periods of time and 77 percent for others. At times when the yield is at the lower value the bonders will be a bottleneck.

One obvious way to relieve the bottleneck would be to improve the yield. Another would be to reduce the C/O time. Changeovers are about 65 percent of the UPtime or OEE detractor, so reducing C/O time would improve UPtime and, therefore, effective capacity significantly. Changing the scheduling methodology to run more logical product sequences (see Chapter 12) would also reduce total C/O time and improve effective capacity.

The reliability of all equipment in this process is quite good, with everything at 90 percent or higher, and everything except forming at 95 to 98 percent. This would suggest that appropriate TPM (total productive maintenance—see Chapter 6) practices are likely being followed. Still, sheet forming should be examined to make sure that best practices are being followed, and to uncover opportunities for further improvement. Although 90 percent is good, the corollary is that 10 percent of the available time is being wasted.

Figure 5.4 shows that sheet forming also has high C/O time and high C/O loss. The loss of two rolls per changeover is probably the more significant factor of the two. With a utilization of only 80 percent, the one hour loss per changeover can be tolerated; it is the cost of the two rolls that has the most influence on the campaign length and the resulting nine-day production cycle (EPEI). The two-roll loss is also a significant contributor to the 13 percent yield loss. Thus, it should be a high priority to reduce the loss at restart; doing so will reduce the waste of defects (yield loss) and the waste of overproduction (long campaigns). It will also reduce the waste of inventory, because the nine-day EPEI leads to the high inventory (12.6 days) between forming and bonding. Solutions to be examined would include reductions in test lab response time, in-line instrumentation to give immediate feedback on properties, and adaptive process control to shorten the time to reach in-spec properties.

It can be seen from Figure 5.5 that there are four forming machines, four bonders, three slitters, and three chopping machines. This is an obvious candidate for cellular manufacturing (explained in detail in Chapter 11). Products should be analyzed to see if they fall into groups or families with similar processing conditions (forming width and basis weight, bonding temperatures, and so on), so that a single forming

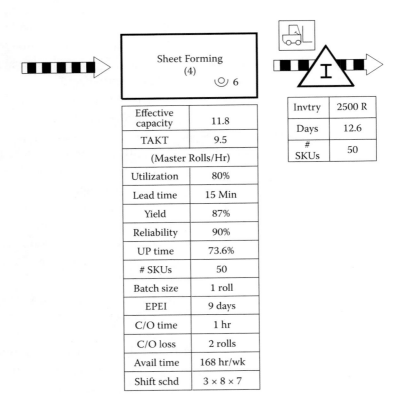

FIGURE 5.4
Material lost in changeovers.

machine, bonder, and slitter could be dedicated to a product family to simplify changeovers, improve yield, reduce inventories, and simplify flow patterns.

LEARNING FROM INFORMATION FLOW

We saw earlier that the information flow portion of the VSM led to a better understanding of why the raw material inventory is so high. This is a clear example of the benefit of showing material flow and information flow on the same map, and why Toyota chose to do so when it started to map its processes.

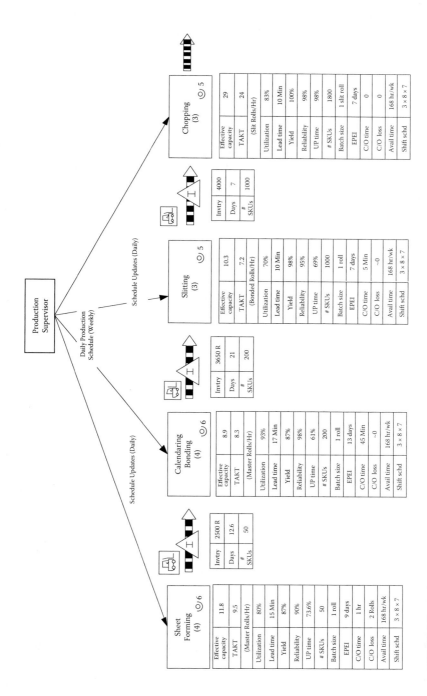

FIGURE 5.5
Opportunity for cellular manufacturing.

Many of the things to be learned from the information flow are triggered by trying to understand reasons for waste in the material flow. Additional questions to be asked include:

- How are customer orders processed? Are there delays in the order entry process? Do these delays hold up material flow and cause inventory? Do these delays prevent a MTO or FTO strategy?
- Is the predominant strategy currently being followed MTS, MTO, or FTO? Is that the most appropriate strategy? Would manufacturing lead times allow FTO or MTO?

As an example, the information flow portion of the VSM created in Chapter 4 (Figure 4.12) shows that this is a MTS operation, is forecast driven, with monthly production based on the next month's forecast minus current inventory. The information flow confirms what the cross-hatched flow arrows in the material flow show—that this is a pure push operation. Moving to a pull replenishment system (Chapter 14) would likely reduce or eliminate many of the wastes we have seen from the VSM analysis.

There can also be waste in the information processing itself, so it is also quite beneficial to stand back and focus on the information flow to see what additional waste may be there. Among the things to be analyzed are the following:

- Are the planning and scheduling processes based on current information or on information that has been batched and queued and is no longer current?
- As the information being processed is transformed, is the result transmitted directly to those who use it, or through other NVA nodes? Do all who need the information receive it? Is it received by people who do not need it?
- Do customers receive appropriate feedback on order status? Are customers consulted for their future needs? Do the demand management process and the forecasting process make appropriate use of it?
- Is forecast information used to reset supermarket levels at appropriate intervals?

TOOLS TO GET TO ROOT CAUSE

While most of the significant wastes in the current process will be apparent from an examination of the VSM, the root causes will not always be readily understood. In those cases, additional analysis tools are employed. The most common are discussed in the following sections.

The Five Whys (5W)

Another of the techniques that Toyota developed, 5W is a deceptively simple but effective way to get to root cause. According to Taiichi Ohno, "The basis of Toyota's scientific approach is to ask *why* five times whenever we find a problem.... By repeating why five times, the nature of the problem as well as its solution becomes clear." Ohno attributes this procedure to Toyoda Sakichi, the original founder of Toyota Industries.

Figure 5.6 gives an example of the application of 5W. In this case, *why* was asked six times, demonstrating that the number "5" is not sacred, that the question *why* should be asked enough times to get to the root of the problem.

Example of 5 Whys
1 Why is the inventory at this location so high?
Because we run long campaigns.
2 Why are the campaigns so long?
Because changeovers are so expensive.
3 Why are changeovers so expensive?
Because it takes a long time to get properties within spec after a restart.
4 Why does it take so long to get back within spec?
Because the relationship between basis weight and pump pressure varies over time.
5 Why does it vary?
Because material builds up in the nozzles and constricts flow.
6 What causes the buildup?
It is the nature of the viscous materials used in this product.

FIGURE 5.6
The five whys.

Detailed Process Mapping

In order to get to root cause, it is sometimes helpful to create a flow map of one area of the process in much more detail than is typically shown on a VSM.

The Ishikawa Diagram

This is a diagram aimed at clarifying cause and effect, named after its creator, Professor Kaoru Ishikawa. It is also known as a cause-and-effect diagram, for obvious reasons, or a fishbone diagram because of its appearance. A complete description of Ishikawa diagrams can be found in any text on quality tools.

Cross-Functional Process Mapping

Sometimes with complex planning and scheduling processes, the information flow portion of the VSM will not allow for diagnosis of root causes of flow problems, or even show all the waste in the information handling processes. In these situations, we have found that the tools described by Rummler and Brache in *Improving Performance—How to Manage the White Space on the Organization Chart* to be effective. Their process includes the development of three maps:

1. A relationship map, a high level diagram showing all the functional entities that participate in or influence the process being analyzed, and the most significant communications and interactions between them.
2. A high level process map in block diagram or flowchart form, to clarify the main steps in the information handling process.
3. A more detailed cross-functional process map, sometimes called a swim-lane chart. This is a two-dimensional view of all interactions, with functional entities down the left margin, and all detailed process steps mapped out horizontally beside the entity that executes them. Figure 5.7 is an example of a cross-functional process map for an order-entry process.

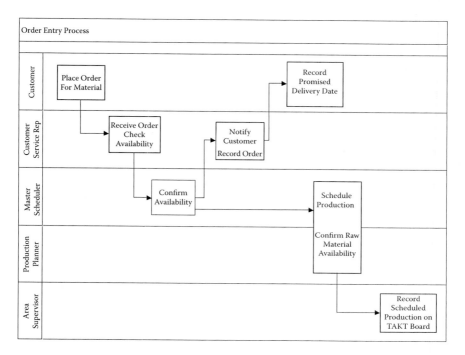

FIGURE 5.7
Example of a cross-functional process map.

If done properly, these maps can depict the waste and the root cause within the information managing processes. They will clarify:

- Where information handoffs occur in a way that information may get delayed or lost
- Situations where several people are involved in a decision, thus adding time, but where one person has all the information and could be given authority to make the decision
- Cases where a single person is involved, but where the information quality or accuracy would benefit from participation and perspectives of others
- Information rework, error correction
- If the cross-functional map is plotted to timescale, delays in information processing will be obvious

CREATING THE FUTURE STATE VSM

Once a thorough analysis of the current state VSM has been done, all major wastes have been identified, root causes have been diagnosed, and potential improvement programs listed, you are ready to organize the improvement efforts into a future state VSM. The future state VSM may be a single document, if all identified improvements are to be implemented simultaneously. If they are to be done in some sequence over a period of time, it is useful to create several generations of future state map, to depict the process status at the completion of each group of improvements. Thus, the future state VSM is not a single vision; it may have several phases, starting with a slightly improved near-term future state and tracing the planned evolution through several generations toward an ideal future state.

There are two distinctly different approaches in the development of the future state VSM, and each provides an additive perspective. The first is to view the current state value stream to envision what a completely waste-free process would look like, one that continuously delivers defect-free products to customers in the quantity they order, delivered when they want it. This creates a view of what an ideal state value stream would be, and is useful in setting an overall vision and direction. The second approach starts from where you are now and looks at specific wastes, root causes, and improvement opportunities to develop a roadmap on how to implement specific improvements to move in the direction of a better future state. So the first approach creates a sense of what a perfect process would be and sets a long-term vision, and the second develops specific plans to get there from where you are. Thus, the specific steps are:

1. Create an ideal state VSM by asking, "What would a perfect, completely waste-free process look like?"
2. Develop a series of future states to move toward the ideal.
   a. Prioritize all improvement opportunities that were identified from an analysis of the current state VSM
   b. Create a future state VSM based on the premise that the highest priority opportunities identified have been successfully implemented

    c. Create a second-generation future state VSM, based on successful implementation of the next set of improvements

    d. Repeat until the future state comes as close to the ideal state as current process technology will allow

Be aware that the ideal future state is not static, that it may change with time as business conditions, customers, products being offered, and other factors change, and as process and equipment technology changes. That doesn't mean that the prior ideal state was not useful; there should always be a vision of the ideal that all current activities are driving toward.

All generations of the future state VSM should be broadly defined before any significant improvement activity is begun. That will help to ensure that a specific improvement is part of an integrated plan and will indeed move the operation toward the desired final future state. And as the ideal future state evolves the maps should be updated to ensure that current activity is still aligned with new direction.

Whether done as a multigenerational plan or as one integrated phase, the various improvements should be sequenced in approximately the same order as they are presented in the later chapters of this book:

1. Implement TPM to improve equipment reliability.
2. Use SMED and other techniques to reduce time and losses incurred in product changeovers.
3. Begin to put visual management practices in place.
4. Find, improve, and manage bottlenecks, areas where the current process can't produce to takt.
5. Implement virtual work cells, if the asset footprint lends itself to cells.
6. Level production flow using product wheels.
7. Consider MTO and FTO as alternatives to MTS.
8. Implement pull replenishment systems, with optimized supermarket inventory levels based on sound mathematical principles.

If done in that order, each improvement is building a solid foundation for the next improvement. If a multigenerational future state VSM depicts various future states being achieved in that order, they have the highest chance of being reached and of being sustained. The later chapters of this book, specifically the chapters on cellular manufacturing, postponement and FTO, production leveling and product wheels, and pull will illustrate various iterations, or generations, of the future state VSM.

Once future state VSMs have been developed, they are not complete unless the data boxes have been updated with the new values to be expected in the future state. Inventories may be lower, lead times may be shorter, changeover times may be shorter, OEE and its other components, yield and reliability, may be higher. It is important to make sure that these are updated for two reasons:

1. To keep performance goals in front of the improvement teams, so they know what level of improvement is expected.
2. To quantify the financial benefits expected. This helps to keep business leaders engaged and motivated, and can be used to justify any required expenditures.

If the future state is a significant departure from the current state, if it has different material flow paths due to virtual cells, has different scheduling methods to level production, or calls for significantly lower inventory levels, there may be some anxiety and hesitation about moving forward. A computer simulation model can be developed to ease the anxiety and build confidence in the plan. Although not perfect predictors of performance, discrete event simulation models allow for reasonable approximations of system behavior and performance to be demonstrated to concerned stakeholders. With on-screen animation capable of showing material processing and movement at rates of hundreds or thousands of times faster than real time, observers can get a much more realistic intuitive feel for the dynamic behavior of the process. These models also allow for "what if ...?" testing of alternate flow control logic, product wheel times, and inventory levels to optimize system performance. FlexSim and Promodel are tools that have seen widespread use in the process industries for this purpose.

The future state VSM should be created by the same cross-functional team that was involved in defining the current state VSM, to maintain the continuity of the improvement process. If the original team had the appropriate membership—that is, representatives of all key stakeholder groups—the vision the team creates for the future of the operation has a much greater likelihood of being widely accepted and becoming a reality.

SUMMARY

A thorough analysis of the current state VSM should be done to understand key detractors to current performance before jumping to creation of a future state VSM. Performance in the future state will generally be much better if the more important of the current problems have been resolved and the process stabilized.

Not all the problems uncovered by the current state VSM will or should be corrected immediately, but all should be analyzed for their impact on future performance and integrated into the future state VSM.

Specific corrective actions and improvement tools required to move toward the future state(s) and, ultimately, the ideal state are described in the remaining chapters of this book.

Part III

Lean Tools Needing Little Modification

6

Total Productive Maintenance

Total productive maintenance (TPM) is a philosophy, a set of principles, and specific practices aimed at improving manufacturing performance by improving the way that equipment is maintained. It was developed in Japan in the 1960s and 1970s, based on preventive maintenance and productive maintenance (PM) practices developed in the United States. But where PM is focused on the maintenance shop and on mechanics, TPM is team based and involves all parts and all levels of the organization, including supervisors, plant managers, and perhaps most importantly, operators. It drives toward autonomous maintenance, where the majority of maintenance is done by those closest to the equipment, the operators. In this operating model, the maintenance group can now focus on equipment modifications and enhancements to improve reliability.

The goal of TPM could be described as the development of robust, stable value streams by maximizing overall equipment effectiveness (OEE).

Some key elements of TPM are:

- **Preventive maintenance:** Time-based maintenance, maintenance done on a schedule designed to prevent breakdowns before they can occur
- **Predictive maintenance:** Condition-based maintenance, using instruments and sensors to try to anticipate when equipment is about to break down so that it can be fixed before failing
- **Breakdown maintenance:** Repairing the equipment after a breakdown occurs
- **Corrective maintenance:** Ongoing modifications to the equipment to reduce the frequency of breakdowns and make them easier to repair

- **Maintenance prevention:** Design equipment that rarely breaks down and is easy to repair when it does fail
- **Autonomous maintenance:** Team-based maintenance done primarily by plant floor operators

The most fundamental element of TPM is the culture change required, moving from a mind-set that the maintenance group owns accountability for equipment performance to one where everyone in the plant has that accountability.

The "productive" aspect of TPM is that all of the preceding should be done in a way that is economical and effective.

It is not the intent of this chapter to explore TPM in great depth, but to highlight its relationship and synergy with lean and its greater value when applied to the process industries.

TPM AND LEAN SYNERGY

TPM should play a prominent role in any lean manufacturing initiative. Poor equipment reliability can result in several kinds of waste:

- The need to protect throughput against failures creates a need for inventory.
- The creation of that inventory is overproduction.
- The need to stock repair parts creates another form of inventory.
- Equipment downtime results in waiting waste.
- Movement waste occurs when maintenance personnel go to the failed equipment, to the spare parts supply, back to the equipment, and then back to the shop.
- Depending on the nature of the failure, defects, scrap, and yield losses might be created by the faulty equipment before the problem is recognized; yield losses are also created after the repair, while the equipment is ramping back to operating conditions and properties are coming within specification limits.

TPM shares with lean a strong focus on teamwork and operator involvement, and the need for vigorous, visible commitment and leadership from upper management. TPM is greatly helped by lean 5S activities:

- Clean equipment runs better; keeping cooling fans clean, and rotating or sliding parts free of dirt and grime, helps prevent accelerated deterioration failures.
- A clean environment makes it easier to see evidence of impending failures: oil spots on the floor; dust from deteriorating drive belts; puddles of process fluids from leaking pumps; and so on.
- A well-organized workplace has the appropriate tools within the operator's immediate reach, thus facilitating autonomous maintenance.

TPM addresses the eighth waste—the waste of human potential—by involving operators in the maintaining process, thereby tapping into their powers of observation and cognitive reasoning. Both TPM and lean rely heavily on visual management for their success, to engage operators more completely in the manufacturing process by enabling them to see more clearly how the process is performing and why.

The net result of this synergy is that TPM is a key part of most lean initiatives. In fact, some companies in the process industries have corporate-wide "uptime excellence" or "maintenance excellence" programs driven by senior leadership. Some have even used TPM as the banner under which all lean improvement work is done. Some use TPM as the name of their production systems.

TPM IN THE PROCESS INDUSTRIES

Asset productivity is far more important to the effectiveness of a process plant than is labor productivity. Therefore, the process industries have an even greater need to do everything within reason to improve the reliability and uptime of the equipment and thus a greater need for the benefits that TPM can bring.

And because equipment is more often the bottleneck to increased flow than labor is, equipment breakdowns generally result in a loss of throughput and, therefore, revenue. Any time that asset utilization reaches into the 90 to 95 percent range, achieving high equipment reliability becomes critical. In the words of *TPM for Every Operator,* "The busier you are, the more you need TPM."

TPM AND RELIABILITY-CENTERED MAINTENANCE

Note that Reliability Centered Maintenance (RCM) is another maintenance improvement process, which grew out of work done for the U.S. Air
Force in the 1950s and 1960s. There is synergy between TPM and RCM
in that their goals and benefits largely overlap; however, RCM is generally
viewed as belonging to the maintenance function or department, whereas
TPM is owned by all.

THE BENEFITS OF TPM

Traditionally there has been a view that reliability and availability could
be increased by increasing the maintenance budget. TPM breaks that paradigm by enabling reliability and availability to be raised without budget
increases. In fact, TPM frequently reduces overall maintenance cost, even
while increasing equipment uptime. Because operators are now doing all
routine maintenance, fewer mechanics may be needed. With less frequent
breakdowns, fewer repair parts are consumed.

The higher equipment uptime resulting from an effective TPM process
enables smoother, more stable material flow with fewer hold-ups and,
therefore, shorter lead times. Pull replenishment systems, discussed in
Chapter 14, become easier to implement. In fact, all the forms of waste
mentioned earlier in this chapter are reduced.

TPM MEASURES

One of the most widespread measures used to gauge the effectiveness of
a TPM effort is OEE. One reason for its popularity is that it captures in a
single metric all the factors that detract from optimum equipment performance. A similar measure used by some companies in the process industries is UPtime. Although the calculation is different, the same factors are
included, and the numerical result is the same as the OEE value. Both OEE
and UPtime are discussed in the following sections.

Overall Equipment Effectiveness

OEE is the product of three factors:

- Availability
- Performance
- Quality

Availability

Availability captures all downtime losses, including breakdown maintenance, preventative maintenance, and time spent in setup or changeover. Availability is calculated as actual operating time divided by planned production time.

Note that setup time or changeover time should not include the time getting properties back within specification after the changeover, because that loss is captured in the quality factor in OEE.

$$\text{Availability} = \frac{\text{Actual Operating Time}}{\text{Planned Operating Time}}$$

Performance

Performance captures the loss in productivity if equipment must be run at less than the design throughput rate because of some equipment defect. For example, chemical batches can take longer to heat up or react if residue has built up on vessel walls, thus impeding heat transfer. Rotating machinery, paper winding equipment, or plastic film processing equipment, for example, may have to be run at slower speeds if bearings are worn. Performance is calculated as actual throughput divided by rated throughput.

A caution on performance calculation: In the process industries, there are often rate limitations due to the requirements of the material being processed, not due to any equipment defect. For example, when heat-treating sheet goods to set properties in, some products may require the heat treater be run slower to allow for more time at temperature than required for other products. Likewise, some batches of paint resin may require more "cook time" to completely react compared to other resins run on that equipment. Because these rate limitations are due to product requirements rather than equipment performance, they are truly value-adding, so the performance metric should not be penalized.

$$\text{Performance} = \frac{\text{Actual Throughput}}{\text{Rated Throughput}}$$

(Total throughput should be a weighted average based on the actual product mix, for equipment that must run at different rates for different products.)

Quality

Quality captures the loss in equipment productivity when out-of-specification product is being made, including scrap material, material that must be reworked to be acceptable, and yield losses during startup or when coming back from a product changeover.

$$\text{Quality} = \frac{\text{Quantity of First Grade Material}}{\text{Total Quantity Produced}}$$

OEE is then calculated as:

$$\text{OEE} = \text{Availability} \times \text{Performance} \times \text{Quality}$$

Figure 6.1 diagrams the buildup of OEE from its component factors.

UPtime

An alternative measure of equipment performance and the effectiveness of maintenance processes is UPtime. Although not nearly as widespread as OEE, UPtime is used by some process companies to drive their reliability improvement efforts. One Fortune 100 process company has standardized on UPtime as a cornerstone of its TPM processes.

(One of the more common uses of the term UPtime is in performance measurement of IT systems, web sites, servers, and other computer networks. However, that industry uses a different calculation for UPtime, which should not be confused with this usage.)

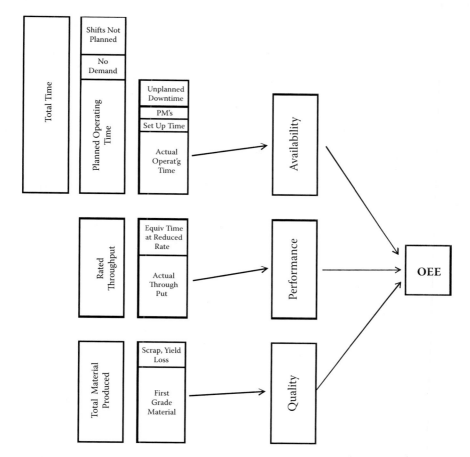

FIGURE 6.1
Components of the OEE calculation.

UPtime captures all the same losses that OEE captures, but calculated in a slightly different way.

$$\text{UPtime} = \frac{\text{Valuable Operating Time}}{\text{Valuable Operating Time} + \text{Losses}}$$

Losses include:

- Unplanned downtime (usually due to breakdown maintenance).
- Preventative maintenance time, including time for major shutdowns and overhauls.

- Setup time, changeover time, transition time, or whatever this is called in your plant. Usually this is taken to be the time from the last good material to the first material made after the transition that is completely within specifications. However, because the waste material made coming up from the transition is included in yield loss below, that portion of the transition or setup time should not be included here.
- Equivalent time at reduced rate. If the process step is running at 80 percent of rated throughput because of equipment problems, 20 percent of the planned operating time would be the equivalent time lost due to reduced rate.
- Scrap and yield loss includes the normal off-spec material routinely made, plus the material wasted in changeovers.

UPtime, then, is the fraction of planned operating time that the process step is able to run at full rate making first-grade product, as shown diagrammatically in Figure 6.2.

UPtime = Valuable Operating Time ÷ Planned Operating Time

UPtime = Valuable Operating Time ÷ Valuable Operating Time + Losses

Calculation of OEE and UPtime

The calculations for OEE and for UPtime give the same numerical results, even though they are calculated differently. As an example of the calculations, consider a plastic film or paper slitting operation, similar to a step in the process described in Chapters 2 and 4.

Calculation of OEE

The relevant data and the calculation of OEE are:

- The plant is operated twenty-four hours, five days per week.
- Demand for products slit on this machine doesn't fully utilize it, so there are twelve hours per week where it is down for "no demand."
- Thus, planned operating time is $(24 \times 5) - 12 = 108$ hours/week.
- There are, on average, two hours per week of unplanned downtime.

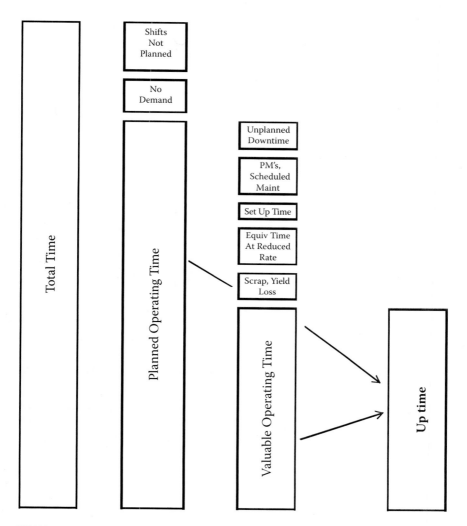

FIGURE 6.2
Components of the UPtime calculation.

- When starting the equipment on Monday morning, two hours of preventive maintenance and equipment cleaning are done.
- This slitter runs sixty different cutting patterns each week. It takes five minutes to move the cutting knives each time, so total setup time is 60×5 min = 5 hours/week.
- Actual operating time is $108 - 2 - 2 - 5 = 99$ hours/week.
- Availability = 99 hours ÷ 108 hours = 91.7%.

- Bearings on the slitter are worn, so if it is run at the design rate, 500 feet/min, the film will move back and forth across the width of the machine, creating edge nonuniformity. If run at 400 feet/min, performance is acceptable. Thus, performance (Actual Throughput ÷ Rated Throughput) = 80 %.
- Approximately 2 percent of the width of the film is lost as edge trim, so yield loss is 2 percent. Quality = 98 %.
- **OEE** = 91.7% × 80% × 98% = **71.9%**

Calculation of UPtime

The calculation of UPtime for the same operation and parameters is:

- The plant is not in operation on the weekends, and twelve hours are not needed due to "no demand," so planned operating time = 108 hours/week.
- Unplanned downtime, breakdown maintenance = 2 hours/week.
- Monday morning PMs = 2 hours/week.
- Setup time = 5 hour/week.
- The equivalent time at reduced rate = 20 percent of the remaining time, = 20% × 99 = 19.8 hours/week.
- The 2 percent yield loss due to edge trim is equivalent to losing 2 percent of the remaining 79.2 hours/week: 2% × 79.2 hours = 1.58 hours/week.
- So the combined losses included in the UPtime calculation are 2 + 2 + 5 + 19.8 + 1.58 = 30.38 hours/week.
- Valuable operating time is 108 − 30.38 = 77.62 hours/week.
- **UPtime** = 77.62 ÷ 108 = **71.9%.**

Thus, the two formulas give the same result. Where OEE is based on a multiplication of percentages, and UPtime is based on subtracting lost hours from total time before taking the final percentage, the two are based on the same combination of losses.

VSM Data Boxes: OEE or UPtime

It should be obvious that the current value of either OEE or UPtime should be included in the VSM data box for each process step. Because a major function of a VSM is to highlight areas of waste in the process, and these metrics capture several equipment-related wastes, its inclusion is critical to full understanding of waste sources. To further that understanding, the components of OEE (availability, performance, and quality) should also be shown. If UPtime is the preferred metric, reliability or a total downtime factor and the yield factor should be shown. It is often helpful to show the setup or changeover time as a separate factor, because that by itself drives waste by encouraging longer campaigns and, thus, more inventory.

In what is considered one of the landmark books on TPM, *Introduction to TPM,* Seiichi Nakajima lists six big equipment losses:

1. Equipment failure
2. Setup and adjustment
3. Idling and minor stoppages
4. Reduced speed
5. Process defects
6. Reduced yield

Both OEE and UPtime capture all of Nakajima's six big losses.

Another value of the OEE or UPtime metric is that it gives operators, who are responsible for all of the routine equipment maintenance under TPM, an indication of how well they are meeting that responsibility. Therefore, the preferred metric should be a prominent part of any visual management activity.

SUMMARY

Proper equipment maintenance is important to the performance of any manufacturing operation, and especially in process plants. Process equipment tends to be very expensive, which means that providing excess capacity (to mitigate equipment downtime) is not often an economic option; therefore, equipment downtime might prevent you from meeting takt.

TPM is a set of principles and practices that have proven to be effective in improving maintenance and, more importantly, improving overall equipment performance. TPM is a process that involves all levels of the organization in the total maintaining process and puts direct responsibility for much of the maintenance work in the hands of those closest to the equipment: the operators.

TPM should be a key part of your lean effort because it can reduce the several forms of waste associated with poor equipment performance: overproduction, inventory, defects, transportation, and waiting.

OEE and UPtime are two metrics that measure overall equipment performance and provide an indication of the effectiveness of a TPM program.

7

Setup Reduction and SMED

For any manufacturing step that must process a variety of product types or materials, determining the total campaign length for each product type is a key operations management decision. If product changeovers are long or costly, there is a tendency to run long campaigns before changing to the next product, to minimize the overall penalty incurred with changeovers. (Note that in this discussion, setup, changeover, and product transition will all be used to denote product changes.)

However, this creates several types of waste. The most obvious is over-production, because more product is now being produced than is currently needed for the next process step. This creates inventory waste, typically both WIP and finished product inventory. Movement waste may be created if the extra material must be conveyed to a remote storage location instead of flowing to the next step. Yield losses may be increased if any of the material is out of specification; it may take longer to get to the point in the process where defects are recognized, so more out of spec material will be produced before the problem is discovered. Perhaps the greatest drawback with long campaigns is that they make the manufacturing process less flexible, less able to react to changes in customer needs.

SMED AND ITS ORIGINS

For all these reasons, it is critical that product changes be accomplished as quickly as possible, so that short campaigns are feasible. Toyota recognized this in the early 1950s as the Toyota Production System was beginning to evolve. One of the most time-consuming changeovers Toyota faced was the replacement of the dies on the large presses used to stamp out auto body parts, which was taking several hours. Shigeo Shingo, an industrial engineer consulting with Toyota, developed a methodology for examining all setup operations and modifying the setup process to reduce the overall time. Using Shingo's techniques, Toyota was able to shorten the die changes from three hours to fifteen minutes by 1962, and to an average of three minutes by 1971. In recognition of this tremendous accomplishment, Shingo's methods and techniques have become the standard for changeover reduction and have come to be known by the acronym SMED, for single minute exchange of dies.

After a brief discussion of SMED concepts, the remainder of this chapter focuses on unique product transition challenges found in the process industries, and additional techniques that can be incorporated into the SMED methodology to deal with them. For a more detailed examination of traditional SMED methods, see *A Revolution in Manufacturing: The SMED System,* which describes SMED from the point of view of Shingo himself.

SMED CONCEPTS

Four of the fundamental ideas that SMED promotes are shown graphically in Figure 7.1 and discussed as:

1. **Move external tasks outside of the changeover time:** Recognize that some of the tasks normally done during a changeover can be done before the equipment is turned off and production stopped. These are called "external" tasks and include things like bringing all the necessary tools and new parts to the equipment. As the setup is nearing completion, moving any parts or assemblies removed from the equipment, housekeeping, and cleanup are tasks that can often be performed after the equipment is turned back on. These external tasks can consume a lot of time so moving them outside of the time window when the machine is not producing can shorten setup time dramatically.

2. **Determine whether any of the internal tasks can be modified:** Determine whether any of the internal tasks can be modified so that

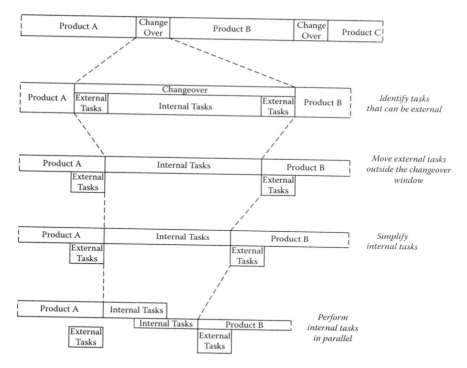

FIGURE 7.1
Major SMED improvement steps.

they are done as external tasks, such as preassembly of any apparatus
or tooling required, and any required preheating of new components
that could be done before being installed on the equipment.

3. **Simplify the remaining internal tasks:** Use dowels, locating pins,
fixtures, and visual marks to speed the time required to get new parts
in place. Standardize bolts where possible to minimize the number
of wrenches required. Use quick-disconnect fasteners where possible. Use poka-yoke (mistake-proofing) techniques to ensure that
apparatus cannot be installed incorrectly.

4. **Where feasible, perform internal tasks in parallel:** If several operators can perform tasks concurrently, the time can be reduced without increasing the total labor content of the setup.

Developing a detailed process map and timing diagram is a good way
to start a SMED activity. Video recording also provides a valuable view

of what actually happens during the changeover. After the changeover process has been revised and tested, it is critical that it be documented, standardized, and audited on an ongoing basis so that the improvements can be sustained.

It is also critical that campaign length then be reexamined, and shortened to take advantage of the changeover time reduction, so that all the wastes just described can be reduced.

The current state VSM provides insight into where SMED can have the biggest benefit. This is not necessarily the steps with the longest changeover times, but the steps where changeover times cause long campaigns (long EPEI), thus creating inventory waste and steps with large changeover losses.

Potential SMED improvements should be shown on the future state VSM, as shorter changeover times, as shorter EPEI, as reduced inventory, and perhaps as reduced changeover losses.

PRODUCT TRANSITIONS IN THE PROCESS INDUSTRIES

In assembly plants, setups generally consist of mechanical and/or electrical modifications to the equipment, subsequent calibration and adjustment steps, and often creation of a test part to check dimensions against acceptable tolerances. In the process industries, we also see tasks of this nature, such as resetting the width of the die in a sheet casting process or changing the extrusion die shape in a breakfast cereal extrusion process. However, it is frequently the case that more of the time is spent in cleaning out the raw material feed systems and the processing equipment to prevent cross-contamination. In many food processing plants, for example, equipment is shared among several product varieties, which may or may not contain allergens, such as peanuts. This can pose very stringent requirements for cleaning between products. As another example, the tinting tanks used in paint manufacturing require thorough cleaning during color changes. The tasks performed during these cleanups are well suited to conventional SMED analysis.

In extrusion, sheet good, and batch chemical processes, much of the time lost is the time required to bring the line to the appropriate temperature, pressure, speed, or thickness, after all the mechanical tasks have been performed, as shown in Figure 7.2. Therefore, the SMED process must also

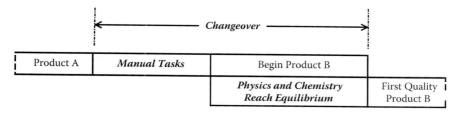

FIGURE 7.2
Components of a process industry changeover.

deal with these so that the total time for the changeover including the time for process conditions to stabilize is reduced. Because these components of the transition are more dependent on the process chemistry or physics than on manual tasks, more technology-related solutions are often employed. These may include techniques like adaptive process control to speed up ramping back to first-grade product and, thus, also reduce the accompanying yield losses. Because changes from one product to another on process lines frequently require much more than mechanical setups, they are often called changeovers.

Consider three different changeover situations typical of process lines, as discussed in the following sections.

A Changeover Where All Tasks Are Completely Manual

Changing knife positions on one of the slitters in the sheet goods process described in Chapter 2 falls in this category. When a new roll with a different cutting pattern is placed on the slitter, all that needs to be done is to relocate the rotating knives to the new positions along their shaft. Once that operation, requiring less than five minutes, is done, the machine can be restarted. SMED could be used to examine the specific steps in loosening the knives, repositioning them, and retightening them to decide whether a different locking technique would be appropriate, or if a second operator working in parallel would speed up the task. If there is any possibility that knives can be positioned incorrectly, consider a more positive positioning mechanism, involving detents, pins, or precise markings. Poka-yoke mistake-proofing techniques could be employed.

Changing the size of the bag or box in a packaging line is another changeover with only manual tasks: cleaning out pneumatic lines; loading the new bag stock; and performing mechanical settings and adjustments.

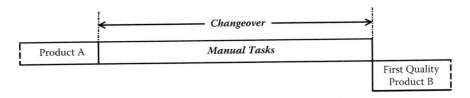

FIGURE 7.3
A changeover with only manual tasks.

Conventional SMED methodology can have great value in improving these setups. Figure 7.3 depicts this type of change.

A Changeover Completely in Chemistry and/or Physics

The temperature on a bonder (heat treater) in our sheet goods process must be changed for a different product type. This is the only task to be done, but it can take considerable time, perhaps an hour or more. The heat treating done by the bonder occurs as the sheet is contacted by a large heated roll. The roll surface has been machined to precise tolerances to give uniform heat transfer at high speeds. When the roll temperature must be changed even by a few degrees, it must be done gradually to prevent warping the roll surface. SMED in this situation may include a structured brainstorming workshop, with mechanical engineers, physicists, mechanics, and operators, to conceive practical techniques for more rapid heating and cooling.

In a food processing plant, products being baked as they are conveyed through an oven may require different oven temperatures or different belt speeds for different products. Thus, the only time-consuming change to be made during the transition is to raise or lower oven temperature; belt speed changes can be done quickly. So the SMED process should include people who are knowledgeable in heating and ventilating and in heat transfer, and in the physics and chemistry of baking processes. Figure 7.4 describes this type of change.

A Changeover That Includes a Combination of Manual Tasks and Chemistry/Physics

Consider the application of photosensitive emulsion in the manufacture of X-ray films. The base film for X-ray products is typically cast and wound onto wide (eight- to twelve-foot) rolls, several feet in diameter. At some

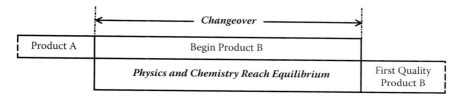

FIGURE 7.4
A changeover where only chemistry or physics changes take time.

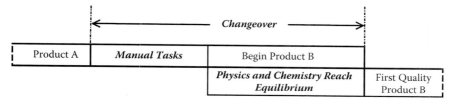

FIGURE 7.5
A changeover with both chemistry or physics and manual tasks.

later time, these cast rolls are removed from storage, unwound and moved through a coating operation, where the emulsion flows through a long narrow die lip onto the base film. The coated film is then rewound and stored for later slitting, chopping, and packaging. The specific emulsion differs by product type and end use: The photosensitive characteristics of films used to X-ray welded structures to detect cracks are quite different from the properties required to X-ray teeth and gums. When product changes occur, the emulsion type must be changed, and the die lip removed and cleaned.

Once these manual tasks have been completed and the line restarted, the film must be run for enough time to allow the emulsion flow to stabilize. Then samples must be collected and taken to a lab to be tested to ensure that the photographic properties are within specifications. The lab procedure involves exposing the film under carefully controlled conditions, developing the film, and then testing its properties. This procedure can take several hours, and during that time the film being produced must be put on "quality hold." If any properties are out of spec, the roll must be scrapped. Only after receiving acceptable results can the product be released as first grade. Figure 7.5 depicts this type of change.

A SMED activity would, of course, address the die replacement and cleaning operations. It may even be suggested that a second die be purchased so that a clean die is always ready to install; the cleaning operation then becomes external to the changeover.

Because most of the changeover time depends on lead time through the lab, SMED would also focus specifically on lab operations. What may then be discovered is that the lab has been managing performance to optimize cost. This is often found in process industry plants, where the lab supervisor's accountabilities would typically be:

- Total cost of lab operations
- Quality/accuracy/repeatability of test results
- UPtime of test instruments
- Development of improved test methods

The lab supervisor's performance rating would generally have little connection to flow or lead time in the manufacturing process.

When this is found to be the case, SMED focuses sharply on flow through the lab and on lab scheduling processes. It may be appropriate to prepare a VSM of the testing lab, including, of course, the information flow. Opportunities would include improving information flow between the coater area and the lab, prioritizing testing sequence, providing technicians for lunch relief, increased staffing during critical periods, and perhaps buying more analyzers to do more testing in parallel. It may be that samples are batched—that is, gathered into groups before testing, causing additional delays. In that case, single piece flow should be evaluated.

Coating experts should also be included in the SMED process, to examine causes of quality variation and to try to improve the physics of the coating application. If coating uniformity can be improved, the time to get to first-grade film will be shortened. Eventually, it may be determined that the loading on the test lab can be reduced.

SMED BEYOND PRODUCT CHANGES

Although SMED can be quite valuable in optimizing product transitions, it has additional uses in process plants. In many process operations, equipment must be changed not because of a product change but because some part of the equipment has become fouled, corroded, eroded, or spent. Even in completely continuous chemical processes that produce a single product 24 hours a day, 365 days per year and, therefore, never undergo a product change, the equipment must be taken off line periodically to replace a catalyst bed, replace a corroded part, or clean residue off of vessel walls. In extrusion processes, including plastic film casting, fiber spinning, plastic pellet forming, and cereal dough extrusion, the extrusion head, die, or spinnerette can become constricted because of accumulated material and require cleaning or replacement well before a product change is scheduled.

If not managed well, and executed in the minimum possible time, these changes incur most of the waste that product changes do. Inventory waste is needed to protect flow against the potential outage. Bringing tools and replacement parts to the equipment creates movement waste. Yield losses can follow the replacement as the process is getting back within acceptable performance limits. SMED, therefore, has value in analyzing and optimizing all these nonproduct change-related tasks. In fact, that is the primary application of SMED in many process plants.

A NON-MANUFACTURING EXAMPLE

Many of the books and articles on SMED use a racecar pit crew as an example of a setup done well. This has been used so often that it has become a cliché, but the reason that clichés become clichés is that they often have a high degree of truth and relevance, and the pit crew analogy has both. Anyone who has watched a professional automobile race can relate to pit crew operations, to the precision, the coordination, and the purposefulness of everything that's being done. The pit crew operation also provides a very strong visual image of SMED principles at work:

- All tasks that could possibly be done externally are. The tools are ready, the new tires are in place, and everything is prepared for the moment when the car enters the pit.
- All tasks have been thoroughly analyzed, simplified, and structured to be done in the shortest possible time.
- All tasks are done in parallel. All four tires are replaced simultaneously, while the gas tank is being filled.
- Technology has been appropriately applied, for example, to the mechanism used to lift the car for tire changes.

- Everyone understands his or her role, and has practiced it to get the time down to the absolute minimum.
- All pit stops continue to be timed, and there is an intense ongoing effort to further reduce time in the pit.

Everyone involved understands that these races are often won or lost in the pits, and is highly motivated to contribute to the potential victory. The more that manufacturing teams understand that operational success is likewise dependent on fast, effective changeovers, the easier it will be for the operation to become lean.

SUMMARY

Most of the equipment found in a process plant must produce a variety of materials in a relatively short time. In some cases, the changeover from one product to the next is fast and inexpensive. In other cases, the changeover can be long and/or costly, driving schedulers to longer production campaigns to minimize the overall transition cost. Long campaigns, of course, create large inventories, which is waste from a lean perspective. To shorten changeovers and thereby promote shorter campaigns, Shigeo Shingo developed an effective work practice that has come to be known as SMED, which has four components:

1. Identify tasks currently done during the changeover that could be done while the process is running, that could be moved to be done before or after the changeover (external tasks).
2. Move the external tasks to be done before or after the changeover. Examine all internal tasks to see whether any of them could be performed externally.
3. Simplify the remaining internal tasks.
4. Perform internal tasks in parallel, where feasible.

These steps are effective at reducing the time required for all mechanical tasks and adjustments required by the changeover. However, many process steps require an additional set of changeover issues to get the product back to first-grade specifications after the changeover. These require that SMED broaden its focus to include:

- Shortening test lab response time
- Providing inline instrumentation to test quality
- Applying improved process control to reduce variation and to reach aim conditions more quickly
- Improving the process chemistry and/or physics to reduce variation and reach aim conditions more quickly

Some process lines have no product changes to deal with, but must be taken off line periodically—perhaps annually—for cleaning and overhaul, and therefore experience similar issues to those found in product changeovers. SMED has proven to be quite effective in improving these periodic shutdowns.

8

Visual Management

One of the most important attributes of a lean manufacturing operation is that it is managed at the locations where the value-adding work is actually done, using visual indications of requirements, status, successes, difficulties, and corrections to be taken. This is contrasted with the more traditional management done remotely, in a conference room or office, using paper or electronic production and status reports. Management in the work areas is more immediate, timely, and responsive; the severity of issues is minimized by shortening the feedback loop.

INTRODUCTION TO THE VISUAL PLANT

Visual management includes six critical elements:

1. A **clean,** visual, well organized **work area**
2. **Basic visual displays,** that define the area and the functions being performed, not unlike a "you are here" map in a shopping mall, but taking that idea further (think about the overhead signs in a supermarket telling you which aisles have bread, soups, and breakfast cereals)
3. **Visual schedules,** that convey to everyone in the area what is to be made, how much, and when, and capture performance to the current plan
4. **Andons and displays of relevant metrics** that further define the condition and status of the operation
5. **Management** of the operation using the information visually conveyed through the displays

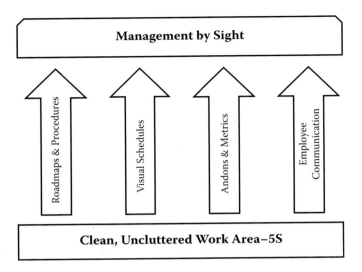

FIGURE 8.1
Components of visual management.

6. Frequent, ongoing **communication with operators, mechanics, technicians,** and anyone else whose assignment places them in the operating area

Figure 8.1 illustrates the relationship between these elements. A clean, uncluttered area is the foundation for all visual management practices. Signs and displays that identify the area, specific pieces of equipment, flow paths, well-identified locations for temporary carts or bins, and placards describing standard procedures are essential in defining the area and the work. Visual display of the current task for each piece of equipment, and progress toward completion give an immediate sense of the heartbeat of the process. Andons indicating equipment status and any problems, and performance metrics and ongoing operator communication provide the remaining components so that the operation can be managed on the floor, at the process, entirely by what can be seen and heard.

Attributes of a manufacturing process with good visual tools:

- Problems can be readily observed and responded to.
- Employees in the area have an up-to-date sense of whether they are on schedule, ahead, or behind. If behind, they can immediately take whatever measures that are at their disposal to catch up. If ahead, they can take action to avoid overproduction.

- Area management has an ongoing sense of production to plan, and can act to bring additional resources or to reallocate current resources when appropriate.

VISUAL WORK AREA

Ohno defines visual management as management by sight, where any abnormal situation can be readily seen and responded to. A key Toyota principle is that no defects or problems are hidden, and a visual work environment will enable this to be achieved.

An effective way to achieve and maintain a visible, visual work area is through a process called 5S. The 5Ss come from the Japanese words *seiri, seiton, seiso, seiketsu,* and *shitsuke,* which are specific steps to address housekeeping, cleanliness, workplace organization, and standard work. They have been approximately translated into five English language terms that, conveniently, also all begin with the letter S.

Specifically, the 5Ss in terms of their English equivalent are as follows:

1. **Sort (*seiri*):** The idea here is that to be visual, an area has to be free of all clutter and nonessential material, that the ideal workplace has only the material and tools required to perform necessary tasks. So this S is about getting rid of all the "stuff" that can get in the way of performing tasks and of line-of-sight management. A common way to accomplish this step is to go into an area and red-tag all nonessential materials. The red-tagged material can then be discarded, or taken to another area if needed there.

2. **Straighten/set (*seiton*):** Arrange everything that remains in the most logical place to enable smooth work flow. Everything should be as close to the point of use as possible, to avoid walking, reaching, bending, or twisting to get to it. Everything should be marked. If tools are hung on a pegboard, outlines should be painted so it is obvious which tool goes where, and if any tool is missing. If tools are on a horizontal surface, foam cutouts perform the same function. Carts that contain parts required for quick changeover should have their location outlined on the floor.

3. **Shine/scrub (*seiso*):** Clean the entire area. Sometimes this includes painting floors and walls a light color, to make drips and spills obvious. A clean work area provides a more pleasant environment, enhances morale, and improves safety.
4. **Standardize (*seiketsu*):** Develop procedures to make maintaining a clean environment part of the standard work for that area.
5. **Sustain (*shutsuke*):** Develop processes to make sure that things don't revert to a messy, cluttered state. Checklists and frequent audits are an effective way to accomplish this. A process plant making thin films and circuit board substrates has been an outstanding example of 5S; the lean team conducts an audit of each area twice a month, and posts the results in run chart format on a wall in the area.

The distinction between the fourth and fifth S, standardize and sustain, may seem somewhat fuzzy; in fact, some books label these systematize and standardize, respectively. The terminology is not really important; what is important is that there is a mechanism in place to make the cleanliness, orderliness, and workplace organization practices inherent to 5S evergreen.

When done well, 5S becomes the foundation for all visual management practices:

- Areas are more clutter-free.
- Flow is more visible.
- Issues and problems are more obvious.

Because workplace *safety* is often a key reason for instituting a 5S program, with the understanding that a highly visual, clutter-free workplace is generally much safer, several process companies have designated their 5S programs as 5S+1, or as 6S.

VISUAL DISPLAYS

People entering a production area should be able to know immediately where they are, what part of the process they're seeing, what function each piece of equipment performs, and how material flows to it and from it. This is especially important in process industry plants, where material is often hidden from view, where all tanks and vessels may look alike, and where liquid, powder, or pellet flow may be in a complex array of pipes.

The things that should be readily seen include:

- Signs to identify the manufacturing area (such as "Resin Production," "Tinting Area")
- Signs to identify each piece of equipment (such as "Resin Reactor")
- Signs to identify the virtual cell to which a piece of equipment belongs (such as "Cell 3 Resin Reactor")
- Descriptions of standard work practices posted near work locations
- Flowcharts posted in the area to show how material flows into this area, and where it goes when it leaves
- Lines painted on the floor to indicate material flow paths
- Tags, signs, and/or color coding on piping to indicate its function and direction of flow
- Signs to designate hazardous areas, and if safety glasses, ear protection, safety shoes, or other protective gear is required

Aids to enable people to understand the geography of the process are an important component of a visual workplace, as are posted procedures.

VISUAL SCHEDULING

The current production task and progress toward it must be displayed in order to be able to manage an operation from what can readily be seen in the area. One of the most powerful ways to accomplish this is through the use of display boards, showing the product being made, the takt rate, production thus far in the campaign compared to takt, reasons for being behind or ahead of schedule, and any corrective action planned. The next products on the schedule and when the changeovers are to be done should also be visible.

Figure 8.2 shows an example of one of these display boards, or takt boards, as some plants have named them, for one of the sheet forming machines in the process described in Chapters 2 and 4. Production for

Production Schedule–Sheet Forming Machine #4

Day	Monday						Tuesday						Wednesday	
Time	12 AM	4 AM	8 AM	12 PM	4 PM	8 PM	12 AM	4 AM	8 AM	12 PM	4 PM	8 PM	12 AM	4 AM
Product	432A					432B				432D			516F	
TAKT (Rolls)	10	10	10	10	10	10	10	10	10	10	10	10	10	10
Produced (Rolls)	11	12	9	8	9	10	9	11	11	10	8	12	10	10
Ahead of TAKT	x	x						x	x			x		
Even with TAKT						x				x			x	x
Behind TAKT			x	x	x		x				x			
Reasons	Changeover went fast				Tough changeover		Paper tear-out				Drive failed	Catching up		
		Ran well	Adjust to TAKT	Adjust to TAKT				Catching up	Catching up				Catching up	

FIGURE 8.2

Example of a visual schedule board, or TAKT board.

several days is shown in four-hour blocks. The specific product type to be made and the takt quantity for that time block are listed. Actual production, and whether it was greater than, equal to, or less than takt is plotted. Reasons for producing more or less than the takt quantity are listed, so that corrective action can be taken for chronic problems. This chart shows that a transition to product 432A was completed on Monday morning at 12 a.m., that it went unusually well and that the machine was running well, so production exceeded takt for the first two time periods. Therefore, an adjustment was made, and production was deliberately lower than takt for the next two periods to avoid overproduction on this campaign. A transition to 432B was done at 4 p.m., and didn't go well so production fell behind takt. Because of that, and a sheet tear out during the midnight to 4 a.m. time period, extra rolls were produced during the later portion of this campaign to catch up to the takt requirement.

A board like this does the following, and is an effective tool for managing the operation.

- Allows all the operators to know what product is to be produced
- Captures production quantity and how it compares to the schedule
- Allows recording of reasons for deviations from plan
- Provides guidance on when to speed up or slow down the process to level production to takt

The takt board is usually accompanied by a board or an easel with a flip chart pad to list all problems discovered, what follow-up action is to be taken, who is to do it, and when it is to be completed. Figure 8.3 gives an example.

With boards like these, shift turnovers become fast, straightforward discussions. The outgoing operator simply reviews the takt board and the action register with the incoming operator, and after a brief discussion of any important details, they're done. Area supervisors can manage much more effectively by checking the takt board and action register, and discussing any unusual events with the operator than by studying reports in a remote office. In many traditional process plants, the primary feedback to supervision occurs in a start-of-shift rack-up meeting. Management by sight shortens the feedback loop dramatically.

For takt boards to be useful and effective, the operators who are the primary users must be involved in their design. The best boards are ones that

Action Register~Tuesday June 2
Wobbling on the #3 idler roll–Fred
Replace idler roll bearings–Ralph–noon today
Drive on web winder failed
Replace drive–Wendy–done
Scraping noise when windup rolls transfer–Fred
Check windup mechanism–Ralph–next product change–wed
Grease spot on floor below unwind roll–Jack
Find source of grease–Ralph–before shift end today

FIGURE 8.3
Example of an action register.

have been designed by a team including operators, supervisors, mechanics, lab technicians, production schedulers, and anyone else who touches that part of the process. For complete acceptance, all shifts should be represented. This is an excellent candidate for a kaizen event (see Chapter 9), and will likely be a short one, taking only a day or two for design and implementation.

A balance must be struck between local board design and plant-wide standards. For maximum utility, acceptance, and ownership, boards should be designed on an area-by-area basis by the people working in each area. However, to make them useful to plant management and to facilitate roll-up of key metrics, there should be plant-wide standards and perhaps templates to guide specific design, so that some degree of commonality is achieved.

The visual boards should be located in the manufacturing area so that they can be kept up to date by operators and mechanics. In process plants, boards are sometimes located in the control room; this works well if the central control room functions as the "nerve center" of the operation, if it is where everyone goes to understand current status, the next tasks to be performed, and potential problem areas. A key principle is that there should be only one board for a specific portion of the process, and that it be in the most accessible location. There might be a tendency to duplicate boards so that the information is more readily available to all, but experience shows that one will often be out of date with respect to the other. It can take considerable effort (waste) to keep them synchronized.

Electronic displays are sometimes used for these functions. Today's display technology allows for inexpensive, large screen displays to be located in the area. Duplicate displays driven from the same computer allow the information to be shared across wider areas and still be completely in synchronization. However, some feel that electronic displays are somewhat impersonal, and don't achieve the same level of operator engagement and commitment that a handwritten board does.

Some plant managers become enamored with visual displays, and feel that more is better. With the best of intentions, displays can proliferate without purpose. All displays must be created for a reason, and have a true purpose, not being done just for the sake of having more visual displays. Displays that convey no useful information go against the principles of 5S, and should be "red tagged" and removed.

Some of the requirements for good visual controls:

- Must be relevant to the operation.
- Must be clear and understandable to all users.
- Should make maximum use of symbols and graphics to minimize text.
- Must be updated frequently (somewhere between every hour to every four hours).
- Must form the basis for the managing process.

The most critical items to be included on the production display boards are:

- Time blocks (hourly or semi-hourly) for the next several days
- Product to be made during each time block
- Takt (production goal for each time block)
- Production to plan (the amount actually produced in each block)
- Cumulative status (ahead of schedule, even, or behind schedule)
- Reasons for deviations (positive or negative) from the plan
- Corrective action to be taken
- Person responsible for action
- Target for completion

ANDONS

Andons are another tool commonly found in a visual plant. *Andon* is a Japanese term that originally meant a paper lantern used as a signal light. As used today it refers to a light or set of lights used to signal that a manufacturing line or a part of a process has a problem or has stopped. A set of green, yellow, and red lights is a common configuration, where green signifies that all is normal, yellow that there is a problem requiring attention, and red that the line has stopped. This is, of course, quite similar to the meaning of these colors on traffic signals.

In a lean parts manufacturing or assembly operation, an important andon principle is that any operator is allowed to stop the line whenever a problem or defect is detected; in fact, operators are encouraged to do so. The andon lights then alert others to the fact that the line has stopped, and where the problem causing the stoppage was discovered. Workers idled by the stoppage are then able to go to the problem area to help diagnose the cause and correct it.

In process plants it is not always that simple. Some processes cannot be stopped without severe quality, continuity, and even safety consequences. However, even in those situations, andons can be useful; they can be used to signal that there is a problem even if stopping the process is not a practical option.

With much of the complex equipment found in the process industries, diagnosing that something abnormal is happening and needs action requires an integrated analysis of a number of signals. Fortunately, most of the equipment in these plants is controlled by PLCs (programmable logic controllers) or DCSs (distributed control systems), which can be programmed to perform this diagnosis. In fact, many chemical, extrusion, and food processing operations are designed with an andon mind-set, even when the designer has never heard that term.

METRICS

The takt boards described display some of the most important metrics related to current production, specifically production to takt, deviations, and reasons for them. There are other metrics that are equally important and should be displayed just as prominently. These include quality results and trends, equipment reliability and trends, and the time to accomplish the most recent changeovers compared to the standard. Although some of these are lagging indicators and are not available in real time, they should be displayed as soon as they are available so people know how well performance standards are currently being met. It is also important for people to have a sense for whether performance is improving, stable, or declining, so trend lines are as important as the most recent values.

Metrics can be written on a white board as soon as they are available, or they can be displayed on a computer-driven monitor. The medium is not important; what is critically important is that they be up to date and that they are presented in a way that relates them to each person's specific work tasks.

MANAGEMENT BY SIGHT AND FREQUENT COMMUNICATION

In order for any of the tools and practices described to be completely effective, they must be integrated into the overall plant managing process. The term "visual management" means just that; displays, andons, and metrics are not effective unless they form the foundation for the way the process is managed. These devices also provide a basis for meaningful communication between supervisors and operators, which often leads to additional dialog that might not have happened otherwise. Managing out on the plant floor, using line-of-sight tools and communication with the people doing the value-adding work, can be far more effective than managing in remote offices and in periodic rack-up meetings.

This is in accordance with the Toyota principle of *genchi genbutsu* (go to the place to see the actual situation), also referred to as "go to *gemba*" (the actual place).

PROCESS INDUSTRY CHALLENGES

Line stoppage is a valuable component of visual management in assembly processes, and can be used within process plants, but with limitations. If a ketchup bottling machine is overfilling or underfilling bottles, that can be sensed and the line stopped. But it is much more difficult to sense online if the ketchup viscosity is out of specifications, so the stop-the-line concept would be difficult to implement to prevent continuation of that problem. Still, line stopping can provide a valuable function and should be employed whenever appropriate.

As noted earlier, there should be good visuals telling employees what PPE (personal protective equipment) is needed in the area in general and what additional PPE is required for specific tasks performed within the area. This is especially important in many process plants, because of the nature of the hazardous materials being handled and because of the equipment used. Employees must wear specialized PPE to perform many of the jobs, so visual management includes clear indication of these requirements.

People sometimes reject the notion of visual management for plants found in the process industries. Equipment is often very large, blocking any possibility of line-of-sight visibility. The materials being processed are frequently contained in tanks, silos, pipes, ovens, or otherwise removed from sight. These are minor limitations compared to the benefit that can be had by applying the other components of visual management:

- Workplace cleanliness and organization brought by 5S
- Visual scheduling of the operation, at the operation, in real time
- Andons to indicate problems and things needing response
- Visual metrics
- Communication with management, at the place where the work is being done

As a final thought, everyone's life is made simpler by the many visual displays we encounter daily in carrying out routine tasks:

- The gas gauge in your car
- A traffic light at an intersection
- The battery charge indicator in the tray at the bottom of your laptop screen
- The bars indicating signal strength on your cell phone display
- The electronic "departures" screens at airports
- The scoreboard at a football game
- Signboards announcing "discount days" at gas stations

We use this information to make decisions about our immediate actions. If the thinking that led to the creation of those displays can be brought to the plant floor, it will make life much simpler for those operating, maintaining, and managing the process.

SUMMARY

For lean operations to be most effective, they should be managed at the location where the work is done, at the *gemba*. To facilitate management at the *gemba*, process areas need to be clean and well organized through the use of tools like 5S. This means keeping the area well marked and the equipment visually identified. Visually display the schedule, so everyone can see what is to be made, how much is to be made, when the next change-over is scheduled, and how the area is performing as compared to plan. Such a display is often called a takt board, because you can tell at a glance if you are making takt. Andons make it obvious when a piece of equipment or the entire process experiences an unscheduled stop. Displaying metrics allows everyone to see performance and trends in quality, reliability, and other important parameters. Although line-of-sight visibility is sometimes difficult to achieve in process plants because of the massive equipment often found in these plants, creative use of signs, labels, and color coding can bring a beneficial degree of visibility.

Putting all these things in place facilitates effective, purposeful communication between managers, operators, mechanics, technicians, and anyone else working in the area, and leads to true visual management.

9

Kaizen Events

The word *kaizen,* as used in Japanese manufacturing, can be defined on several levels. It is a term used to signify continuous improvement, which is at the heart of both lean and the Toyota Production System. It can describe a *process* of making small incremental improvements on a daily basis, which ultimately leads to major improvement over time. It can describe an *attitude*, instilled in the workforce, that workers have the responsibility to identify and participate in potential improvements to their work processes. And it can be viewed as described by Jeffrey K. Liker, in *The Toyota Way,* as "a total philosophy that strives for perfection and sustains TPS on a daily basis."

KAIZEN BY SPECIFIC EVENTS

Although kaizen should be practiced every day to achieve incremental improvement, it is often useful to form a team and dedicate a short period of time to resolving a specific problem. Kaizen events are focused, team-based activities aimed at doing specifically that. *Kaizen for the Shopfloor,* by Productivity Press, defines kaizen events as "a team activity aimed at rapid use of lean methods to eliminate production waste in particular areas of the shopfloor. It is well-planned and highly structured to enable quick, focused discovery of root causes and implementation of solutions."

This is an excellent definition, but is somewhat limiting in that kaizen events have been used to create dramatic improvements far outside the shopfloor, to simplify and take time and waste out of:

- Transportation steps in global supply chains
- Accounts receivable processes
- Order entry processes used by customer service representatives (CSRs)
- Patent application and filing processes used by a legal department
- Job posting, interviewing, and candidate selection processes used by a human resources (HR) department

Kaizen events, also called *kaizen blitzes* and *kaizen workshops,* have been described thoroughly and well in several excellent references, including the two just cited, so only the highlights are presented here, along with some cautions and some factors more specific to the process industries.

Kaizen events are of course aimed at making improvements to some defective or wasteful part of the process, but they are also intended to teach specific lean tools and build skill and experience in their application to real problems.

The typical kaizen event is three to five days in length, but can be much shorter; depending on the scope, a kaizen event can be successfully completed in a day or less. One of the fundamental concepts inherent in kaizen events is that the improvement must be successfully demonstrated during the workshop. The improvements are not only conceptualized and designed, but also implemented as part of the event. Thus, the time selected must allow for that, and the event must be scheduled at a time when the process is being performed. A kaizen event may tackle a complex and difficult setup that is performed infrequently. The event must be timed so that the setup is to be done during the event. Observation of a prior occurrence of that setup may be done as part of the event planning.

Kaizens are a powerful way to make improvements quickly and in a way that engages all stakeholders. The benefits include:

- Things get done. Improvements are accomplished; waste is eliminated.
- Results are seen quickly.
- Participants learn lean tools and get experience in application.
- Participants gain a better understanding of other parts of the manufacturing operation, due to the cross-functional nature of kaizen teams.
- The solution is owned by the creators.
- A successful event builds energy and momentum for the entire lean effort.
- Employee motivation and morale are improved.

QUALITY CIRCLES VERSUS KAIZEN EVENTS

Quality circles are another kaizen process that originated in Japan and then gained popularity in the western world in the 1980s and 1990s. Development of the quality circle concept is generally attributed to Kaoru Ishikawa, also known for developing the Ishikawa diagram, or fishbone diagram, as it is also commonly known.

Although kaizen events and quality circles are both team based and aimed at continuous improvement, they are different processes and shouldn't be confused. Key differences include:

- Quality circles are longer-term, ongoing team activities, not focused on a single problem.
- A circle doesn't end with a specific improvement, but takes on new tasks as the previous one is complete.
- They are less structured processes.
- Participation is typically voluntary.
- Quality circles require far less advance planning and preparation; kaizen event preparation steps such as data gathering are usually done as part of the ongoing work of the quality circle team.

Similarities include:

- An underlying philosophy that improvements are best made by the people who actually perform the process tasks.
- Participant training, but quality circles typically focus more on quality tools than on lean tools.
- Quick implementation of improvement ideas, by the team that conceived them.

STEPS IN THE KAIZEN EVENT PROCESS

Selection of appropriate topics for kaizen events is important. Ideally, they should be directed at waste and flow issues found from the value stream map (VSM). They should be improvements that, when implemented, will contribute to movement toward the future state VSM. In short, they should be a component of an integrated plan for overall process improvement. They should not be randomly chosen simply to meet a goal to stage a kaizen event every month or to satisfy a quota to conduct a certain number of kaizens within the next quarter or year. The term "drive-by kaizen" has been used to characterize this unfortunate practice.

Training is a significant part of a kaizen event. It is typically done on a just-in-time basis as needed during the event. It generally includes training on the kaizen workshop process itself so that participants will know what to expect and what is expected of them. An introduction to value stream mapping is presented, leading into discussion of the VSM specific to the process being improved. Then lean tools appropriate to the scope of that event are presented.

A kaizen event involves three major phases: planning; the workshop itself; and the follow-up.

Planning

Although the event itself will take only a few days, the planning must start weeks in advance. This is particularly true for kaizen events in process plants, in that they typically require more planning and up-front data gathering than is needed for events in assembly operations. Specific steps include:

1. Define the scope.
2. Set the objective.
3. Decide on the participants, which typically include those directly involved and those in roles peripheral to the process being improved. Consider adding a wild card, someone not directly involved or familiar with the area being improved, who can ask the naïve, thought-provoking questions. If a multi-shift operation, include representatives of all shifts to ensure that what is implemented becomes standard work across all shifts.
4. Select the kaizen event leader. The most important qualifications are good interpersonal skills and organizational and project management skills; these are more important than experience in the area being improved.

5. Decide on workshop length and design the agenda. Some books prescribe a standard agenda; however, it is preferable to design the agenda based on the specific objective, deliverables, and time frame selected.
6. Schedule the event.
7. Arrange for the participants to be available.
8. Arrange for management to be available at the appropriate times during the event.
9. Define the data and other information that will be required (such as detailed process maps).
10. Schedule the collecting and organizing of the prework information.
11. Arrange suitable facilities (projector, wall space, room for participants to move around). Much of the event will take place at the operation being improved, but there is usually a need for a place for the team to meet in surroundings more suitable for training, map analysis, idea brainstorming, and improvement documentation.
12. If equipment must be taken off-line during the event, arrange for that to be done as part of planning. In some cases, SMED kaizens, for example, the process can continue to run, and the team can observe current state changeovers and try new changeover techniques as part of normal product sequencing. In other cases the equipment must be stopped so that the improvement can be physically implemented. In these cases the interruption to operations must be planned for in advance.

Conducting the Event

Some of the specific details will vary depending on the focus and scope of the event, but the essential elements are:

1. Kick off the workshop with introductions, purpose, scope, ground rules, boundaries, and a management discussion on workshop objectives and the importance to the operation of achieving them.
2. Agree on the improvement target and a way to measure it.
3. Briefly train about the kaizen event process, value stream mapping, and the specific lean tools to be employed (the last may be delayed if an improvement approach has not yet been selected).
4. Study the problem and all available data and maps.

5. Propose a solution.
6. Implement the solution.
7. Fine-tune or modify the solution as necessary to achieve a practical solution that meets the target.
8. Document the solution.
9. Define and document a control plan to sustain and institutionalize the gains.
10. Present the results to management, including the process owner.
11. Celebrate success!

Following-Up

Follow-up activities are as important as the event itself. Without good follow-up, the improvements are not likely to be sustained. The more important tasks include:

1. Communicate and publicize the new standard process to all who perform it and to all support people.
2. Audit the results; track key metrics on an ongoing basis.
3. Look for opportunities to leverage to similar situations elsewhere in the process and anywhere on the plant.

APPROPRIATE EVENT SCOPE AREAS

Kaizen events can be used to address a variety of problems and improvement opportunities. The most common lean tools applied through kaizen events are 5S, SMED, and standard work. It is appropriate to start with relatively simple tools, to introduce the kaizen event concept, gain some quick success, and build momentum for more challenging scopes. For the more complex scopes, it may be more appropriate to manage them as projects, and to use the kaizen event process for specific pieces of the project plan. For example, implementation of pull replenishment on one of the more complex operations found in a process plant may take far more than a week to analyze all the demand variability data, decide where the supermarkets should be located, and calculate the required kanban levels. Additionally, these complex processes touch enough people that it may not be practical to get representatives of all stakeholders together for one event. In that case, managing the entire scope as a project, with kaizen events scheduled to address specific project tasks, such as analyzing demand data and deciding which products will be MTS, FTO, or MTO, achieves all the objectives that make the kaizen event process desirable. In some cases the entire project may be a sequence of kaizen events. In those situations, much of the planning for the full sequence can be done together, with specific outcomes of early events feeding some of the detailed planning of later events. The following manufacturing improvement practices are typically implemented through kaizen events:

- 5S
- Visual management (design and construct takt boards; develop a process for collection and displaying relevant metrics)
- SMED
- Standard work
- VSM development
- TPM pilot
- Yield improvement
- Bottleneck identification and management
- Cell design and implementation
- Product wheel design and implementation
- Pull system design and implementation

KAIZEN DANGERS: THE ROOT CAUSES OF KAIZEN FAILURES

Although all three phases of a kaizen event are important, the planning phase may be the most important of all. Even an extremely well executed event cannot overcome the effects of poor planning.

There are a number of factors involved in a successful kaizen event; a complete list would include mention of each of the steps enumerated earlier. However, there are some critical success factors that rise above the others:

- A well-defined, focused, achievable scope
- Clear boundaries

- All the required data and information identified and collected prior to the event
- Participation by representatives of all the people who perform or touch the process on an ongoing basis (operators, supervisors, mechanics, lab technicians, CSRs, master schedulers, and so on)
- Full-time participation by everyone on the kaizen team
- Management ownership of and participation in the process

Unfortunately, not all kaizen events are completely successful. The reasons why some kaizen events fail or don't live up to their full potential include:

- **Poor planning, not enough planning:** This has been mentioned before, but it is such a prevalent cause of kaizens failing that it needs emphasis.
- **Drive-by kaizens:** Selecting the scope for some arbitrary reason or selecting a scope that is not tied to a clear operational goal or to achieving the desired future state. This practice can cause some degree of immediate success and, therefore, enthusiasm for kaizens, but leads to disappointment later when the collective benefit of all the completed events is small.
- **Too large a scope, taking on more than can be reasonably completed within the time allotted:** The scope of the event must be one that the team can reasonably be expected to complete within the number of days scheduled.
- **Lack of management guidelines and boundaries:** Teams sometimes believe that they have been given more latitude than they actually have, go beyond the level of change that management is willing to support, and then feel disenfranchised when their recommendations are not supported.
- **Distracted participants:** If team members leave the room to answer phone calls or to return e-mail, it weakens the whole process. Their input may be vitally needed while they are out, and they may not be supportive of ideas developed in their absence.

A well-planned and well-executed kaizen event is a beautiful thing to witness, but to someone outside of the kaizen process it may look easy. Kaizen events may be simple, but they're not easy. It takes a lot of hard work to make them successful.

PROCESS INDUSTRY UNIQUE REQUIREMENTS

In most respects, kaizen events conducted in process operations are similar to those in assembly plants. However, there are sometimes differences, although more in degree than in concept.

The processes and the equipment found in process plants tend to be more complex, so the problems addressed by kaizen events may be more complex. For example, designing and implementing a product wheel for a process making sheet goods of various widths, thicknesses, densities, and tensile strength may be more complicated than creating a mixed model schedule and developing a heijunka board for a process assembling gear pumps.

A SMED kaizen must address not only the mechanical changes made to the equipment during the changeover, but also may need to focus on cleaning and decontamination issues, getting the process parameters back to on-aim conditions, and response time from the testing lab. In a food processing plant, where some products contain allergens and others not, changeovers include extensive cleaning and sanitizing. SMED must, therefore, examine novel means to accomplish the cleaning, which may require that equipment designers be included in the kaizen event. Machinery that cuts and bales fibers to go into apparel production needs extensive cleaning so that long fibers from the prior run don't mix with shorter fibers in the current run. Tanks used to tint house paint require thorough cleaning before the next color can be tinted. Kaizen events to improve changeovers on any of these processes must focus on much more than the mechanical tasks involved, and may require participation from plant process engineers.

Process industry kaizen events may require even more planning and data gathering than their assembly counterparts. In assembly processes, *genchi genbutsu* is often an effective way to understand the current state. Assembly operations, inventories, bottlenecks, and flow discontinuities are often readily visible. In a process plant, many of the operations take place within large vessels, ovens, or chambers so that the material being processed can't be seen. Inventories are often stored in automatic storage/ retrieval systems or silos or tanks where the contents can't be estimated by eye. Thus, in many cases, current inventory data cannot be gotten by walking the floor and counting. Therefore, much more thought must be given to deciding which data will likely be needed, as well as much more time in gathering that data.

The number of participants on a process industry kaizen team may be greater than would normally be expected for a kaizen event in an assembly plant. More groups or functions are typically involved; in addition to operators and mechanics, the team frequently includes production planners and schedulers and test lab technicians. If yield is part of the issue, product development specialists may be required. If a changeover on a complex piece of equipment is to be improved, development engineers may participate. There is no real consensus on ideal kaizen team size; Rath and Strong suggest a maximum of eight, *Kaizen for the Shopfloor* recommends six to twelve, while *The Toyota Way* suggests a maximum of fifteen. Process plant kaizen teams can push even that upper limit: the team for a recent kaizen event focused on changeover of a complex casting line consisting of sixteen participants:

(4) Shift operators
(2) Mechanics
(1) First line supervisor
(1) Test lab technician
(1) Area manager
(1) Master scheduler
(1) HR representative
(2) Mechanical development specialists
(1) Product development chemist
(1) Black belt
(1) Lean sensei

Managing team interpersonal dynamics can be a challenge with a team this large, because you're making sure that everyone is treated with respect and as equals, and that everyone has a chance to be heard. But that will not be a problem with a skillful team leader.

On the other hand, if the scope is small, a process plant kaizen team may need only five to eight members.

KAIZEN EVENTS AS SIX SIGMA PROJECTS

If the Six Sigma problem-solving methodology is well ingrained into an operation and becoming a part of the culture, kaizen events should be planned and conducted as Six Sigma projects. The DMAIC framework will increase significantly the likelihood of a successful outcome, and will not slow down or encumber the process if done properly.

The design (D) and measure (M) phases will help to ensure that all the required factors are considered as part of the kaizen planning, that all customer CTQs (critical-to-quality characteristics) have been identified, and that the necessary data and process maps are identified and prepared. During the workshop, the analyze (A) and improve (I) phases can ensure that all normally required steps are considered. Creation of a control plan with the required auditing and follow-up is more likely to be done properly if the control (C) phase framework is followed.

There should be a Black Belt or Green Belt on the kaizen team, and that person should be the team conscience to ensure that the methodology is being appropriately followed, and document the work in the Six Sigma template.

Experience shows that kaizen events managed as Six Sigma projects are more successful and have a far greater likelihood of being sustained. The key is to follow the Six Sigma methodology but not get carried away with it; it should be viewed as a framework to identify all steps that may be required for success, not as a strict roadmap. For example, kaizen events rarely work on measurement system accuracy (MSA), but it is useful to consider the accuracy of your data and treat it accordingly. Six Sigma usually requires a great deal of precision in the data, where lean often drives toward an 80/20 understanding, so when implementing lean as Six Sigma, one must take care to maintain the appropriate balance, and not let a zeal for data precision inhibit analysis and implementation progress.

Although there may be pieces of the Six Sigma process that are not needed in a specific kaizen event, most will be, including:

- Chartering
- Process mapping
- Gathering current state data
- Prioritizing the list of possible root causes of the problem being improved
- Proposing and piloting a solution
- Documenting the solution
- Putting a control plan in place

If the people who will be participating in the kaizen event are already somewhat familiar with these steps from their prior Six Sigma experience, they will already understand many of the key components of the kaizen process.

SUMMARY

Kaizen events are an excellent way to improve manufacturing processes and work practices. The kaizen team includes the people who do the work every day, so that hands-on operational experience is incorporated into proposed solutions. Although kaizen events are completed quickly—usually within a few days—the process includes implementing and testing the improvement, so results are seen quickly. By involving the workforce in these short, focused improvement activities, the organization moves toward a culture of continuous incremental improvement.

Planning the event is fundamentally important to kaizen success. This includes setting the scope to one that provides a meaningful challenge to the team, as well as one that can be completed within the set time limit. Process industry kaizen events often require more planning because much of the necessary data cannot be collected through a *gemba* walk. This means that requirements must be scoped and data collected in advance.

Process industry kaizen events often tackle problems that are more multidimensional than those in assembly processes, so the event team may require people from a wider variety of job functions and, therefore, more participants.

The event itself is usually a high-energy, slightly stressful, yet very gratifying activity. In many ways, it is similar to the two-minute drill at the end of a close football game, where the offensive team is a few points behind. The goal is very clear: score, or the game is lost. The time available is fixed: When the clock runs down, you must have completed the task. You have planned carefully for this situation: It's a page in your playbook.

So you put the best team on the field with the right skill sets to execute the plays chosen, concentrate on making progress in a timely manner, and drive toward success!

Part IV

Lean Tools Needing a Different Approach

10

Finding, Managing, and Improving Bottlenecks

One of the primary objectives of lean is to achieve smooth continuous flow of material through the process. In order to make progress toward this goal, bottleneck resources must be identified, managed, and improved. A bottleneck is any resource whose capacity is less than or equal to the planned throughput, any machine or process step where the takt rate equals or exceeds capacity.

BOTTLENECKS IN PROCESS PLANTS

Bottlenecks tend to have different causes and to have more severe implications in the process industries. In parts manufacture and assembly, people tend to be the rate-limiting factor in many steps, so managing bottlenecks is often a matter of managing people, by appropriate staffing and task leveling. In process plants, throughput in most manufacturing steps is limited by equipment capability, not by labor. In cereal manufacture, throughput can be limited by baking times or by flake extrusion rates; in papermaking by the lineal speed capability of the forming, bonding, or calendaring machinery; and in paint making by the time required to complete the chemical reaction in resin production. With equipment rather than operating labor causing the bottleneck, throughput limitations can't be resolved by bringing in additional labor or by scheduling overtime. And because many process plants run around the clock on a 24/7 schedule, scheduling extra shifts is not the answer. Further, because equipment tends to be expensive and relatively inflexible, replacing or upgrading equipment is not often a viable option. So managing the bottleneck is a matter of optimizing the performance of the bottleneck resource itself, protecting the bottleneck from upstream and downstream problems, and optimizing bottleneck scheduling.

Because OEE factors (see Chapter 6) all have variability, what is not a bottleneck at some times may become one at others. For example, a 90 percent yield value shown on a VSM is an average, and doesn't mean that you lose 10 percent every day. That process step may run at 99 percent yield on a good day and 80 percent on a bad day. The step may have plenty of capacity on the 99 percent day but be a bottleneck on the 80 percent day. Similarly, variability in equipment reliability on a day by day basis can cause non-bottlenecks to become bottlenecks for periods of time.

It is important to recognize that throughput can be limited by factors directly related to a piece of equipment and its performance, either its inherent processing rate, its downtime, capacity lost to yield losses, time lost to changeovers, or all the above. But throughput can also be limited by the manner in which a piece of equipment is scheduled and how well its flow is synchronized with upstream and downstream process steps. In *Synchronous Manufacturing,* Umble and Srikanth explain this distinction, referring to the former as bottleneck resources and the latter as capacity constraint resources (CCRs): A "capacity constraint resource [is] any resource which, if not properly scheduled and managed, is likely to cause the actual flow of product through the plant to deviate from the planned product flow."

As an example of a CCR, consider the case of a cereal plant that manufactures two families of cereal, one formed into thick shapes like stars and circles, and one formed into relatively flat flakes of various shapes. The plant can be divided into three major areas (Figure 10.1), shape manufacturing, flake manufacturing, and packaging, which includes bagging, boxing, cartoning, and palletizing.

From the data boxes contained on a more detailed map, packaging has a utilization of only 75 percent, even though it takes the full output of both cereal production lines. However, in real life the storage silos often became full and forced a production line to go down. Analysis revealed that although the packaging area appeared to have excess capacity, it was being scheduled with no coordination or synchronization with either production area, so it became a constraint.

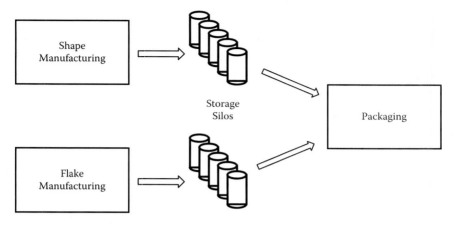

FIGURE 10.1
High-level view of a cereal plant.

Most of the discussion in this chapter on moving bottlenecks, hidden bottlenecks, root causes of bottlenecks, and managing and improving bottlenecks applies to CCRs as well as to bottlenecks.

The following summarizes the characteristics specific to bottlenecks found in process plants:

- The root cause is generally in equipment capacity and performance, not in labor staffing.
- Plants often run around the clock, so additional shifts are not a feasible solution.
- Root causes include yield losses and reliability problems, as well as inherent capacity.
- Nonbottlenecks can become bottlenecks due to variability of OEE factors.
- Bottlenecks may move with product mix.
- Bottlenecks may not be obvious; the resulting waste is usually hidden.

MOVING BOTTLENECKS

Identifying and managing bottlenecks can be difficult in process plants because the bottleneck may be at a different process step for one material

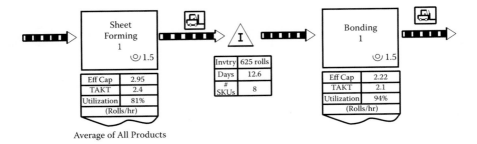

FIGURE 10.2
Forming and bonding utilizations based on average effective capacity.

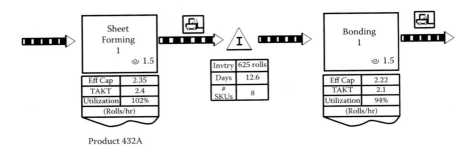

FIGURE 10.3
Product that causes forming to be a bottleneck.

being produced than for another material; the bottleneck may move as the process cycles through the various products being made. As an example, consider one sheet forming and one bonding machine from the process mapped in Chapter 4, shown in Figure 10.2. As can be seen from the data boxes, each has effective capacity greater than the takt requirement, so neither is a bottleneck. However, the values shown in the data boxes represent the averages taken across the full product lineup at the typical mix. When forming sheet with high basis weight, the forming machine must run at much slower lineal speeds, so for that product the capacity will be less than takt and the machine becomes a bottleneck, as depicted in Figure 10.3. With other products, forming may have excess capacity while bonding may become the bottleneck. For products that must be bonded at higher temperature, the line speed must be slower to allow the sheet to be in contact with the heated bonding roll long enough for complete heat transfer from the roll to the sheet. So when making products requiring high bonding temperatures, bonding becomes the bottleneck, as illustrated in Figure 10.4.

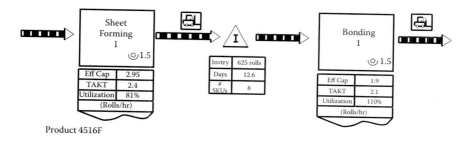

Product 4516F

FIGURE 10.4
Product that causes bonding to be a bottleneck.

In a process spinning synthetic fibers, the threadline winding machine may be the bottleneck when producing fine fibers, while the metering pump feeding the extrusion die may be the bottleneck with thicker fibers.

In a salad dressing bottling line, the bottle filler will usually be the bottleneck operation when filling the larger bottles, but the label applicator can become the bottleneck when packaging into smaller bottles. The carton erection and filling operation may have significant excess capacity when packaging in large cartons for the "big box" discount retailers but become the bottleneck when filling small cartons for convenience stores.

The fact that the bottleneck may move during the production cycle must be recognized so that appropriate bottleneck management strategies can be used with all process steps that can be bottlenecks.

RECOGNIZING COVERT BOTTLENECKS

To manage bottlenecks appropriately, it is necessary first to identify where they exist in the process. One way to find bottlenecks is to look for locations where inventory tends to build up. But keep in mind that inventory buildup can be for reasons other than bottlenecks, for example, storage of materials not being produced at this point in the production cycle. Nonetheless, a large inventory at a point in the process is a clue that the next step might be a bottleneck resource or a CCR.

In many process plants, however, the in-process inventory is not visible; it is stored in a location somewhat removed from the main process flow, so it doesn't give obvious clues to possible bottlenecks. In sheet goods processes, for example, large rolls are generally stored in a rack system out of the main process flow. These systems may store WIP rolls from several points in the process, and may not have specific slot areas dedicated to specific WIP points, so a visual inspection of current rack contents will not give a view of the amount of material awaiting any specific process step. Similarly, portable stainless steel tanks used to store resins in a multi-step paint making operation may be intermixed within the storage area, so a large number of portable tanks in storage will not indicate which process step is causing the hold up. Plastic pellets and cereal flakes are often stored in large silos within the process, again masking any visual indication of large inventory buildup. Electronic storage area management systems can generally provide reports of storage area contents, but it may take some manual sorting and regrouping to understand current storage by WIP location, so diagnosing possible bottlenecks may take some effort.

Computer simulation models of the manufacturing process will help with both the covert inventory issue and the moving bottleneck problem. A discrete event simulation model, using a tool like ProModel or FlexSim, can identify which steps are bottlenecks or near-bottlenecks and how that changes with each specific product in the overall mix.

THE ROOT CAUSES OF BOTTLENECKS

Once it is clearly understood which process steps are bottlenecks or near bottlenecks, the next step is to diagnose the root cause, to understand why that step is a bottleneck. The most common reasons in many process plants will include:

- Inherent equipment capacity limitations
- Long changeovers
- Mechanical reliability problems
- Yield losses
- Inappropriate scheduling (CCRs)

If an apparent bottleneck is not a CCR but a true bottleneck, the root cause can be diagnosed from the data box for that process step. The data box for a resin reactor is shown in Figure 10.5, and with a utilization of

Resin Reactor

Cycle time (Capacity)	50 GAL/MIN
TAKT	55 GAL/MIN
Utilization	110%
Lead time	2 HR
Yield	97%
Reliability	98%
Uptime	74%
# SKUs	12
Batch size	8000 GAL
EPEI	36 HR
C/O time	40 MIN
C/O losses	1%
Available time	168 HR/WK
Shift Sched	12 hr × 2 sh × 7 d
No of operators	1

FIGURE 10.5
Resin reactor data box.

110 percent it is clearly a bottleneck. Regardless of how well it is scheduled, it lacks the inherent capacity to keep up with takt. A closer examination of the data box shows that yield and reliability are both very good, at 97 percent and 98 percent, respectively. The culprit, the primary contributor to the 74 percent UPtime, is the long changeover time. At forty minutes, it currently gets performed twelve times during each thirty-six-hour production cycle, so 22 percent of total capacity is being lost in changeovers. The key to opening up this bottleneck is to reduce changeover time, using the SMED techniques described in Chapter 7.

The manufacture of synthetic fibers used in blends with cotton to make T-shirts and sweatshirts usually concludes with a process step where the fibers are cut to relatively short lengths (½ inch to 1½ inch), and then baled in 500-pound bales similar to bales of cotton. In fact, it is packaged in this form to enable the fabric converter to process it on cotton equipment and, thus, simplify the blending operation. The data box for a cutter/baler used

Fiber Cutter/Baler

Cycle time (Capacity)	30 Bales/Shft
TAKT	35 Bales/Shft
Utilization	117%
Lead time	9.6 Min
Yield	97%
Reliability	70%
Up time	60%
# SKUs	14
Batch size	1 Bale
EPEI	4 Days
C/O time	45 Min
C/O losses	1%
Available time	168 Hrs/Wk
Shift Sched	8 × 3 × 7
No of operators	1

FIGURE 10.6
Fiber cutter/baler data box.

in this type of process is shown in Figure 10.6. With 117 percent utilization, it is clearly a bottleneck. Again, changeover times are a contributor: With fourteen changeovers consuming forty-five minutes each being done in every four-day cycle, 11 percent of total capacity is lost. However, that is not the primary culprit; even if changeovers could be completely eliminated, the baler would still be a bottleneck. Mechanical/electrical reliability of the baler can be seen to be 70 percent, so machine failures cost the operation twenty-eight hours on every ninety-six-hour cycle, compared to the eleven hours lost to changeovers.

If the mechanical reliability could be increased to 90 percent, the total UPtime percentage would rise to 78 percent. Utilization would drop from 117 percent to 90 percent, thus, the bottleneck would be resolved. However, achieving 90 percent reliability on this type of equipment is quite challenging, so eliminating the bottleneck condition will likely require a combination of reliability improvements and changeover time reduction.

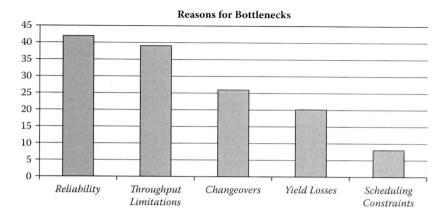

FIGURE 10.7
Reasons for process plant bottlenecks and CCRs.

An unscientific estimate of the relative frequency of causes of bottle-necks and CCRs in process plants can be seen in Figure 10.7. This chart was prepared from a small sampling of VSMs from a variety of process industry manufacturing lines, by examining the data box of the bottle-neck, near-bottleneck, or CCR and estimating the primary cause. When more than one reason was a contributor to the constraint, partial weight was given to each.

BOTTLENECK MANAGEMENT: THEORY OF CONSTRAINTS

Once the bottleneck has been identified and the root cause diagnosed, it's time to open it up, to make whatever changes that can be identified to resolve the bottleneck. At the same time, it is important to make sure that throughput at the bottleneck is not suffering unnecessarily from problems upstream or downstream of the bottleneck.

It may require inventory to accomplish this, but adding that waste is often the most reasonable compromise when compared to the alternative of not being able to make takt. Consider as an example, the process shown in Figure 10.8. The bottleneck is obviously Step B, with a utilization of 104 percent compared to 50 percent for each of A and C. However, if the process is configured exactly as shown, Step B will not be able to process 48 gallons

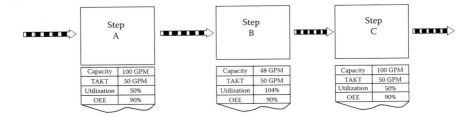

FIGURE 10.8
Flow through an unprotected bottleneck.

per minute (gpm); rather, it will be limited by downtime and losses at A and C, so its real throughput will be only 39 gpm (48 gpm × 90% × 90%).

If buffer inventories can be located as shown in Figure 10.9, Step B will be able to achieve its effective rate of 48 gpm. The bottleneck hasn't been relieved, but at least it is not further constrained by upstream and downstream outages. The contents of the first inventory should be equal to the throughput of the bottleneck multiplied by the longest expected outage of Step A. If Step A can fail and be down for two hours at a time, this inventory should be 5,800 gallons, so that Step B can continue to be fed at its rate of 48 gpm for the entire 120-minute outage. The storage capacity of the second inventory should likewise be determined by the bottleneck rate and the longest outage of Step C. However, this inventory location should be kept empty or nearly so; its function is to provide a place to store bottleneck output while Step C is down, so that outages at Step C will not shut Step B down.

The strategies for managing and optimizing bottlenecks described by Eli Goldratt in his two landmark works, *The Goal* and *Theory of Constraints*, have become the standard for dealing with them. The process laid out by Goldratt can be summarized as:

1. Identify the bottleneck.
2. Exploit the bottleneck. Make sure that the bottleneck is running at maximum capacity, and not wasting time on noncritical tasks.
   - Use SMED techniques to reduce changeover times to the minimum possible. Focus not only on mechanical tasks but also on cleaning, and on getting to specified conditions and properties quickly after the changeover.
   - If the bottleneck is also a CCR, make sure that scheduling processes are coordinated and synchronized to eliminate that portion of the limitation.

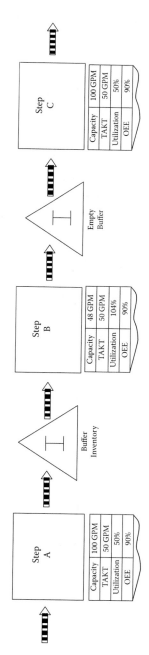

FIGURE 10.9
Protecting the bottleneck.

3. Subordinate everything else to the bottleneck. All upstream and downstream processes should operate in a way that maximizes bottleneck throughput as described.
4. Elevate the bottleneck capacity; try to increase the capacity of the bottleneck.
   - Diagnose yield losses; consider improved process control techniques, both electronic controls and statistical process control (SPC). Six Sigma can also be particularly effective in these situations.
   - If the bottleneck is due to capacity limitations inherent in equipment design, structured brainstorming workshops with mechanical and/or chemical experts can often point to cost-effective remedies.
   - If equipment reliability is the core issue, implement TPM as described in Chapter 6.
5. Once the bottleneck is broken, find the next bottleneck and repeat.

WIDENING THE BOTTLENECK: LURKING BOTTLENECKS

After the most obvious bottleneck has been opened and is no longer a bottleneck, when that process step can now produce to the takt requirement, it should not be taken as given that the process can make takt. There may be other steps that have been bottlenecks, but in a way that was masked by the most restrictive bottleneck. Of course, a complete and accurate VSM would have shown that, but on some process lines the primary bottleneck restricts flow in a way that it is difficult to measure effective throughput at adjacent steps. In other cases, area managers may make assumptions about bottleneck location without the benefit of a good VSM.

Consider the recent experience of the plant manager of a facility that makes salad dressings, mayonnaise, and bottled sauces. On one particular salad dressing bottling line, market demand had increased takt from the previous 300 bottles per minute (bpm) to 400 bpm. The plant manager's intuition, reinforced by some data, told him that the bottle filling step was literally the bottleneck (this is one case where "bottleneck" is more than a figure of speech!). So he challenged his technical organization to increase bottle filling to 400 bpm. After some analysis and preliminary design, they informed the manager that with a redesign of the filling nozzles, the filling operation could indeed be elevated to 400 bpm; however, the line speed would increase to only 320 bpm. What he hadn't seen, and what no one else had seen until the VSM in Figure 10.10 had been developed, was that there were three other near-bottleneck steps that would become bottlenecks at the new takt of 400 bpm. The homogenizer step in the kitchen area would reach its limit at 60 gpm (320 bpm), the label applicator at 360 bpm, and the case packer at 33 cases per minute (400 bpm). Although it would be economically feasible to replace nozzles on the bottle filling machine, the total cost of eliminating all bottlenecks was prohibitive, so the decision was made to build a new line instead of upgrading the current one.

Had the technical team not decided to map the entire process, and had the lurking bottlenecks been overlooked, the plant manager would likely have spent the money on the new nozzles and then realized that the line was still 20 percent short of meeting takt.

SUMMARY

Identifying bottlenecks and potential bottlenecks is important in any operation, but often more so in process manufacturing operations. Many of these lines run around the clock, seven days per week, at or near full capacity, so there is no extra time available to create additional capacity. This means that improving performance of the bottleneck step—and in most process plants you will generally find a bottleneck or near-bottleneck—is critical to meeting customer takt.

Process bottlenecks are just as often due to reliability problems and equipment downtime as they are to inherent capacity limitations, so TPM and OEE improvement programs are even more important to process plants.

Theory of constraints provides an effective process to manage bottlenecks and ensure they aren't further constrained by upstream or downstream outages.

The bottleneck in a process plant may move to a different step with each material being produced, thus complicating bottleneck identification and management.

Managing and opening up bottlenecks can be a less costly alternative to capital projects as a way to increase throughput, so the topic deserves significant attention and focus in capacity-limited process operations. And because effective bottleneck management requires inventory buffers, bottleneck improvement is important even on lines with excess capacity.

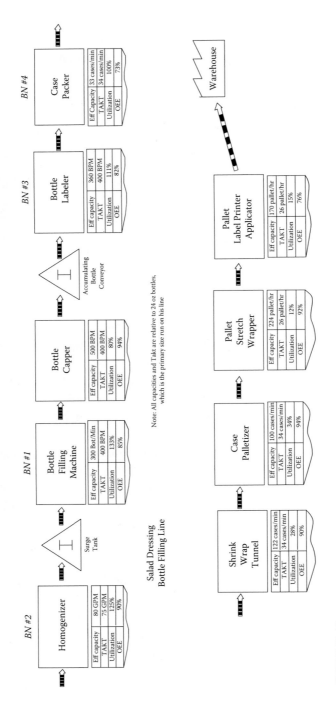

FIGURE 10.10

VSM of a salad dressing bottling line.

11

Cellular Manufacturing in the Process Industries

Of all the improvement tools in the lean toolbox, cellular manufacturing is perhaps one of the most powerful. It enables smaller lot production, more visible flow, quality improvements, reduced WIP, shorter lead times, and simplifies the implementation of pull replenishment systems. However, it has been largely overlooked by lean implementers working in the process industries. There has been a widespread feeling that because of some unique challenges in process plants cellular manufacturing would be difficult to apply. In this chapter it is shown that if we are willing to broaden our understanding of the cellular concept, it is indeed practical to apply to process plants, and has sometimes been the most beneficial lean tool applied.

THE PROCESS LAYOUT (PRE-CELLULAR MANUFACTURING IN ASSEMBLY PLANTS)

An equipment configuration traditionally found in assembly manufacturing plants prior to the advent of lean thinking is depicted in Figure 11.1. In what was called a process layout or a functional layout, all the machines of a given type were grouped together in a shop area dedicated to that type of processing. Thus, there was a shop where all the drilling was done, another where all the lathe operations were performed, and another for grinding.

This layout was attractive for several reasons. Each shop or area could be supervised by someone who had practiced that craft and, therefore,

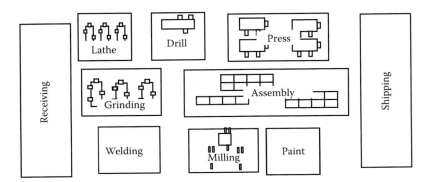

FIGURE 11.1
A functional or process layout.

had the experience to schedule the work and coach and supervise the workers. It was believed that each area could be equipped and staffed for the throughput required for that craft function so that labor productivity could be maximized.

Arranging machines by function, however, created problems. Material would move from shop to shop in batches, transported in tubs, totes, or on carts. To reduce the movement requirements, transportation lot sizes could get large, thus creating a lot of WIP. If production in the various areas wasn't carefully coordinated and synchronized, totes could accumulate in front of each shop, resulting in even more WIP. If the tooling on a machine in one of the shops got worn, and began to produce out-of-tolerance parts, significant time could elapse before a later process discovered the problem, so scrap rates would rise and quality would suffer. There would be little visible sense of flow. Pull replenishment systems would be difficult to implement.

THE PRODUCT LAYOUT (CELLULAR MANUFACTURING IN ASSEMBLY PLANTS)

To overcome these problems, Toyota rearranged equipment into work cells, where each cell would contain all the machines required to process a family of parts. If the machines in a cell were arranged in a U-shaped pattern (Figure 11.2) or an L-shaped pattern (Figure 11.3), one worker could

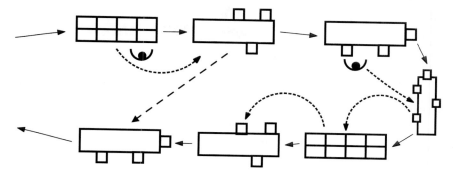

FIGURE 11.2
A U-shaped cell.

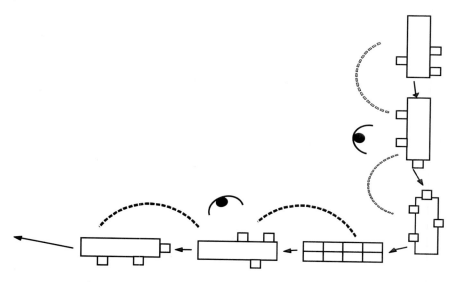

FIGURE 11.3
An L-shaped cell.

operate several machines, so labor productivity could be maintained or enhanced. In Ohno's words "instead of having one worker per machine, one worker oversees many machines, or more accurately, one worker operates many processes. This improves productivity."

Flow through the cell can be in very small lots; one piece flow is often achieved. Flow becomes much more visual and easier to manage and improve. Pull becomes much more practical. Quality is enhanced, because a part processed on one machine is immediately fed to the next operation;

thus, quality feedback is immediate. Because flow within a cell and flow from cell to cell can be better synchronized, WIP is reduced.

There are two key steps in cell design. The first is to group all parts into families, where the parts within a family require similar processing, a step often called group technology (GT). The *APICS Dictionary,* for example, defines group technology as, "an engineering and manufacturing philosophy that identifies the physical similarity of parts (common routing) and establishes their effective production. It provides for rapid retrieval of existing designs and facilitates a cellular layout."

The next step is to create the work cell for each part family by selecting the specific machines required by that family and rearranging them into a cellular configuration similar to those in Figure 11.2 and Figure 11.3.

An additional advantage of cellular manufacturing is that because a given cell processes only a subset of the entire parts catalog, with similar processing requirements, machine setups and changeovers are generally simpler and faster.

CELL APPLICATION IN THE PROCESS INDUSTRIES

Despite the advantages that cellular manufacturing has brought to parts manufacturing and assembly, it hasn't found widespread application in the process industries. Many have concluded that because process equipment is so massive, with so many process interconnections, it would be highly impractical to rearrange into U-shaped or L-shaped patterns. The paper forming machine in the process described in Chapter 2, for example, is approximately 30 feet by 150 feet, weighs many tons, and has process piping and hydraulic and pneumatic lines connected to it. A project to relocate it would cost several million dollars. Similarly, the process vessels used to polymerize resins in paint manufacturing are typically sized for 5,000- to 10,000-gallon batches, have many process interconnections, and would require similar funding to relocate.

In *World Class Manufacturing,* Richard Schonberger describes cellular manufacturing and cites industries that are well suited to this arrangement. He mentions light assembly, such as electronics, and machining industries; he makes no mention of process industries. This may have been appropriate in 1986 when the western world was just beginning to adopt lean concepts and it was important to publicize the applications with the most obvious benefit. What is somewhat disappointing is that during the next 20 years, little was published to expand the view of the applicability of cellular techniques. As recently as 2003, Fawaz Abdullah, a PhD candidate at the University of Pittsburgh, wrote a dissertation that was generally an insightful look at the process industries. However, he had this to say about cellular manufacturing:

> Managers have been hesitant to adopt lean manufacturing tools and techniques in the continuous process industry because of reasons such as high volume and low variety products, large inflexible machines, and the long setup times that characterize the process industry. As an example, it is difficult to use the cellular manufacturing concept in a process facility due to the fact that equipment is large and not easy to move.

A few process companies implementing lean learned years ago that many of the benefits of cellular manufacturing realized in parts manufacture can be achieved without actual equipment relocation. Through the creation of what Wayne Smith calls "virtual work cells," benefits similar to all of those reported for parts manufacture can be realized, with some additional ones.

And others are more recently beginning to write about cellular concepts for process plants. Tom Knight, writing in an Invistics white paper, talks about "flow lines," which is a somewhat similar concept, and Schonberger has spoken recently about "logical cells."

TYPICAL PROCESS PLANT EQUIPMENT CONFIGURATIONS

A generic configuration typical of many process industry plants is depicted in Figure 11.4. There are a small number of key processing steps, in this case four, and there are a few (three to ten) machines, tanks, or reaction vessels in parallel at each step. The parallel machines are quite similar, and often a specific material can be processed by any one of them. Occasionally, the machines or vessels have some unique capabilities such that some materials must go to a specific machine or vessel.

Figure 11.5 shows this configuration for the sheet goods process described in Chapter 2 and mapped in Chapter 4. There are four similar roll-forming machines, four similar roll bonders, three slitting machines,

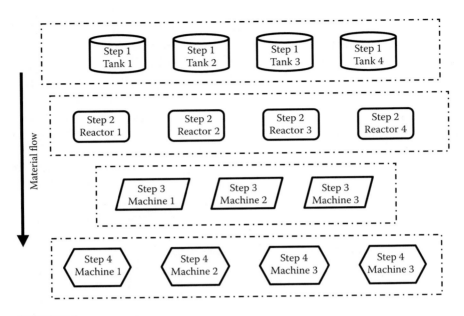

FIGURE 11.4
Typical process industry equipment footprint (functional configuration).

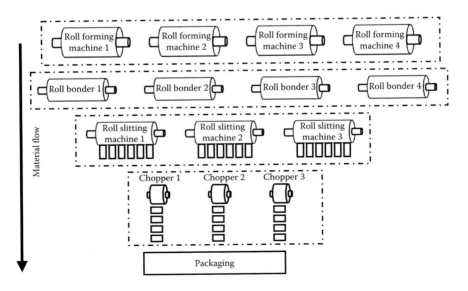

FIGURE 11.5
Sheet process equipment configuration.

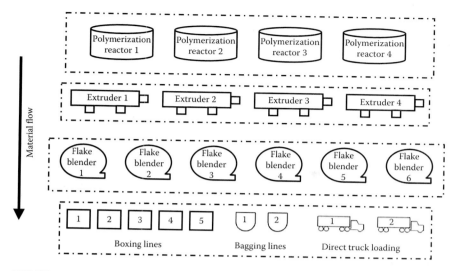

FIGURE 11.6
Plastic pellet manufacturing equipment configuration.

and three cutters or choppers. A roll produced on any of the four forming machines may be processed by any of the bonders, slitters, and choppers.

A similar footprint can be found in a plastic pellet manufacturing operation (Figure 11.6), with four parallel polymerization reactors, four extruders, six flake blenders, and nine packaging lines to box or bag the pellets or load them directly into a hopper truck.

Similar equipment patterns can be found in the manufacture of architectural paints (Figure 11.7).

Process plants usually require this array of equipment to handle the high volume of material to be produced. Practical equipment size limitations prohibit the design of a single machine or vessel large enough to process the full throughput required. Product mix considerations (resulting from the high degree of product variety) would encourage the use of many small machines to give more flexibility, but economies of scale have deterred plant designers from going in that direction. Because capital cost optimization has traditionally overridden good lean thinking in process plant design, the result is a few large vessels or machines at each process step.

This equipment configuration is highly valued for the flexibility it offers. If a batch of material is leaving Step 1, and one of the Step 2 machines is down for maintenance, there may be others available to process the material. The result is that flow paths are often as shown in Figure 11.8. All the flexibility

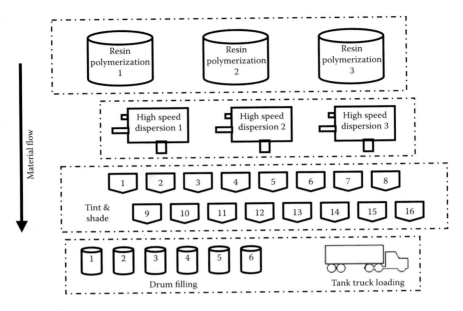

FIGURE 11.7
Paint manufacturing equipment configuration.

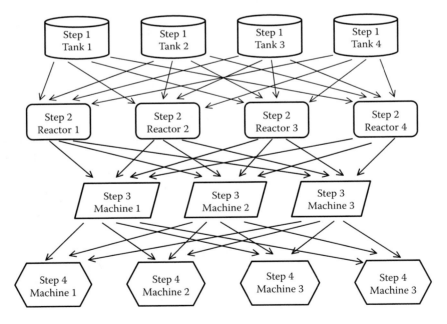

FIGURE 11.8
Typical process industry flow patterns.

inherent in the system is exploited, but generally with more negative than positive consequences. There is frequently a belief that utilizing this flexibility maximizes asset utilization, although the opposite is usually true.

This mode of operation, exploiting the inherent flexibility of this configuration, brings most of the problems described for the traditional functional layout in assembly processes. Material tends not to flow directly from one step to the next, but to be put into some type of storage. In the sheet goods process, formed rolls typically do not flow directly to a bonder: Instead they are taken to an automatic roll storage system, to be retrieved later for transport to a bonder. Thus, large WIP storages are created. Flow becomes unsynchronized, is difficult to visualize, and even more difficult to manage. Pull is extremely difficult to implement.

Quality suffers for two reasons. Like in parts assembly, there is a significant time lapse between each process step, so any defects or out-of-spec material may not be discovered for some time, making all the intermediate material suspect. Even with this simple-looking arrangement, there are 192 (that is, $4 \times 4 \times 3 \times 4$) possible flow path combinations. Because no two machines or vessels will produce exactly the same product, that provides 192 different ways that process variabilities can add up. A statistical process control (SPC) specialist would tell you that you don't have a single process, you have 192 different processes. With so many variables, root cause analysis of product defects becomes difficult.

Because there are thought to be alternate paths available whenever a piece of equipment fails, there is far less urgency to maintain the equipment appropriately. Thus, with time, equipment performance as measured by OEE or UPtime deteriorates.

VIRTUAL CELLS

Cellular manufacturing could overcome most or all of the problems described in the preceding section. However, as noted, because of equipment size and interconnectedness it is impractical to rearrange it into U- or L-shaped arrangements. But if one considers that the primary benefit of the U or L arrangement is that one worker can operate several machines, as a way to optimize labor productivity, and that all the other benefits of cells can be attained without rearrangement, it becomes apparent that a process plant can reap the benefits by managing flow in a cellular fashion,

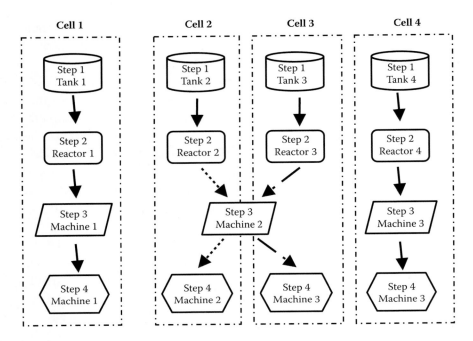

FIGURE 11.9
Grouping into virtual work cells.

without any equipment relocation. The key is to think in terms of *flow* rather than *function*.

The basic concept is similar to that described for assembly processes: Start by grouping all materials into families requiring similar process conditions. Then identify the process equipment required by each family, but instead of creating a work cell by rearranging the equipment, create virtual work cells by defining the acceptable flow patterns. Figure 11.9 shows what this would look like for the process diagrammed in Figure 11.4 and Figure 11.8. Again, no equipment would have to be moved; the new, more limited flow patterns would simply have to be defined and followed.

It should be noted here that although process plants have the logical process configurations shown, they are not generally arranged that way geographically. The equipment is often dispersed in what appear to be illogical arrangements. One reason for that is that many process plants are fifty to one hundred years old and have been reengineered several times as products became obsolete and new ones replaced them. Because process equipment is expensive, there has been a tendency to try to reuse as much of it as possible, to add new equipment only when absolutely necessary, and to locate it wherever it would fit, not where it would facilitate smooth flow. So as plants evolve, the equipment arrangements typically become more and more scattered. The point is that virtual work cell patterns are not as obvious as these diagrams would make them appear.

The advantages of the virtual work cell concept shown in Figure 11.9 are quite similar to those described for assembly processes. Flow becomes far easier to understand, to visualize, to manage. Flow tends to be much more continuous, with far less material being transported to storage, so WIP and material handling are reduced. Quality improves because feedback is much more immediate. As depicted in Figure 11.9, we now have only four possible flow paths instead of the 192 we had before, so product variability is reduced.

It must be recognized that the numbers don't always work out perfectly, but reasonable compromises that will give most of the benefit can usually be found. In the case shown, since there are only three machines at Step 3, one must be shared between Cells 2 and 3. If that machine didn't have enough capacity to process the total throughput of the two cells, the flow shown in Figure 11.10 would be required. But even with this compromise, we have only six path combinations, still far better than the original 192.

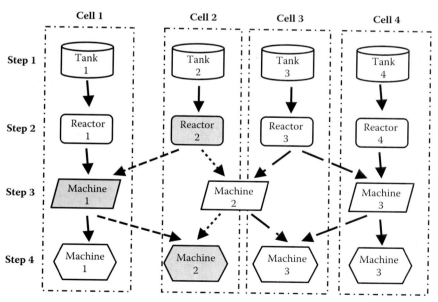

FIGURE 11.10
Alternate virtual work cell grouping.

In the case where a piece of equipment is shared between cells, it is very important to maintain cellular *flow*. The batch of material processed on Reactor 2, and then on Machine 1 at Step 3, should return to Machine 2 for Step 4 processing. It should not be allowed to flow to Machine 1 at Step 4; allowing this to happen will increase the number of possible paths and, thus, begin to reintroduce the quality and flow problems inherent in multiple flow paths. If, for example, all material leaving Step 3 is allowed to take either possible route, the number of flow combinations increases from six to twelve.

It is possible for a cell to include more than one piece of parallel equipment at a process step. If the process shown in Figure 11.10 had six reactors at Step 2 rather than four, the cellular arrangement might look like Figure 11.11.

Figure 11.12 shows a virtual cell layout for our sheet forming process. Because there are only three slitters, Slitter 1 would be shared between Cells 1 and 2. Because the choppers are designed to cut from different incoming slit widths, all three are required by each cell, so the choppers

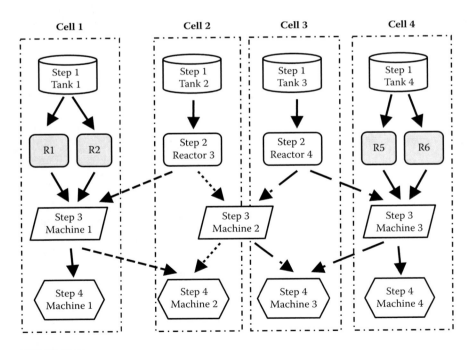

FIGURE 11.11
Alternate virtual work cell grouping.

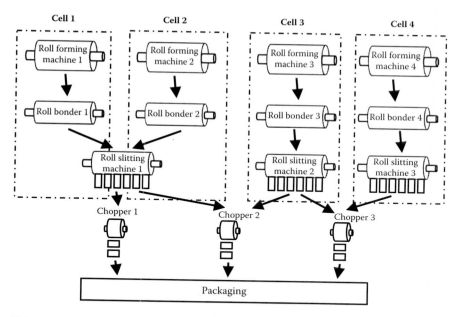

FIGURE 11.12
Sheet process cell grouping.

are not included in the virtual cells. Even with that, the flow is restricted to twelve possible combinations, where prior to the formation of virtual cells, 144 possible paths (that is, $4 \times 4 \times 3 \times 3$) existed. Thus, the key advantages of cellular manufacturing apply.

Case Study: Virtual Cell Implementation in a Synthetic Rubber Production Facility

Cellular manufacturing was applied to a synthetic rubber production process, and worked out more advantageously than usual because a configuration was found where the takt, capacities, and numbers of pieces of process equipment could be almost equally balanced.

Figure 11.13 depicts the process, which begins with three tanks where monomers and other ingredients could be weighed so that the correct quantities could be fed into one of six polymerization kettles. After the batch polymerization was completed, the polymer could be fed to one of three tanks where a chemical process called emulsion stripping could take place. The stripped emulsion was then stored in tanks (not shown) and

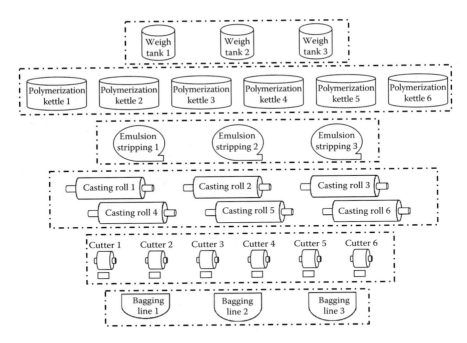

FIGURE 11.13
Synthetic rubber manufacturing configuration.

later cast onto one of six freeze rolls where the emulsion would solidify into a thin sheet. The sheet was then gathered into a rope and cut into pellets, which were bagged on one of three bagging lines, and palletized for shipment to the finished product warehouse.

Customers of this synthetic rubber would use it as received, or as compounded with other materials, to make high pressure hoses, industrial belting, and gaskets and seals for refrigerator doors and car trunk lids.

The product lineup consisted of three major families: type F; type J; and type R. The volume was distributed among the families in the approximate ratio of 45 percent F, 35 percent J, and 20 percent R. Each family comprised eight to fifteen individual grades.

Figure 11.14 is a high level view of the VSM of the rubber manufacturing process.

The pre-cellular scheduling process was:

- A monthly sales and operations planning (S&OP) process would determine the production needs for the coming month, by specific grade within each family.

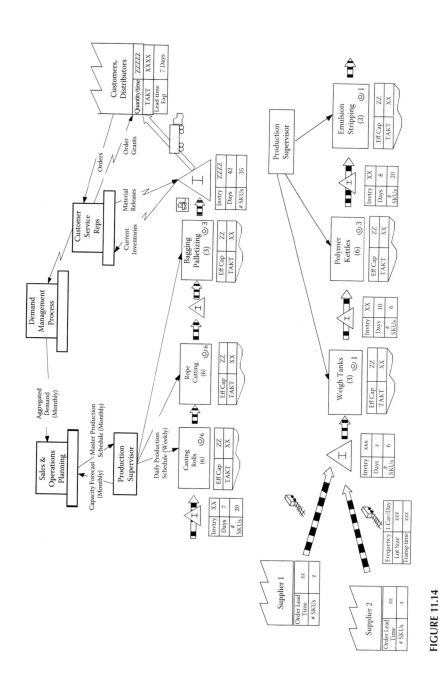

FIGURE 11.14

Synthetic rubber current state VSM.

- The plant production scheduler would determine the grade with the most immediate customer due date, say grade J-43.
- The entire production facility was set up to make grade J-43. Production would continue until the full requirement for J-43 had been produced.
- The scheduler would then determine the grade with the next most immediate customer due date, say F-6.
- The entire facility would be reconfigured to produce F-6. The entire monthly requirement of F-6 would be produced.
- This scheduling process would be repeated until all production requirements for the month had been met.

Operating in this manner created a number of problems:

- The polymer area, the stripping area, the casting and cutting area, and the packaging area were run independently, each with its own area supervisor. There was little coordination between process areas. Consequently, flow was very nonsynchronous. Batches often had to wait for hours in the storage tanks between steps, and often got out of sequence. Asset productivity suffered as a result.
- Because flow patterns were confusing, and because there was no single individual responsible for overall flow, batches occasionally got pumped to the wrong place, and subsequently had to be "ditched" (sent to the waste sewer).
- Large quantities of WIP usually resided in the storage tanks as a result of the flow discontinuities.
- Because each grade was produced only once per month, large quantities of finished product inventory had to be maintained in the warehouse.
- Transitions from one family to another were complex and time-consuming.
- After major transitions were mechanically complete, and the process restarted, it would take several hours for viscosity, the most important product characteristic, to come within customer specifications. This created significant yield losses because of material waste. This time required to reach aim conditions would further deteriorate asset productivity.
- Another result of the flow discontinuity was that a pull replenishment system was virtually impossible to implement.

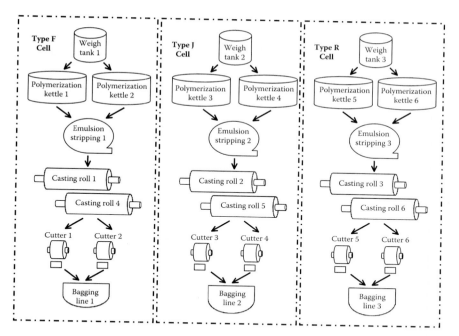

FIGURE 11.15
Synthetic rubber virtual cell configuration.

The Result: Synthetic Rubber Virtual Work Cells

To overcome these problems, a cellular configuration was designed for the rubber process (Figure 11.15). The numbers worked out remarkably well in this case. Because there are three major product families, it seemed logical to create three virtual cells. And because there are either three or six pieces of equipment at each significant process step, dividing the equipment into cells was straightforward.

Even though product demand wasn't equally distributed across families, with Type F seeing 45 percent of the total demand, the productivity improvement resulting from virtual cell implementation enabled even the Type F cell to have enough capacity to produce to takt.

The results of operating in accordance with the cellular plan were dramatic. The most significant benefits resulted from the reduction in yield losses while getting back within viscosity specifications after a changeover. In the old scheme, it would take several hours, perhaps four or more, to get back to aim conditions after a change from F to J, J to R, or R to F. Because the new cells would each produce only grades within a type, target viscosity could be reached rapidly, typically within two to four minutes. This not only improved yield significantly, it improved asset productivity, because the time previously spent getting on aim could now be spent producing first-grade product.

The organizational structure was also changed, to be based on flow rather than on function. Instead of an area supervisor for each process step, there was now a flow manager for each cell, responsible for all the equipment within that cell and for synchronizing flow within the cell.

Specific benefits recorded by the business were:

- Scrapped material was reduced by 28 percent.
- Finished product variability, measured by standard deviation of viscosity of first grade product, was reduced by 15 percent.
- Lead time through the complete process was reduced by 28 percent.
- Average transition time was reduced from eight hours to three hours.
- Average time to reach aim viscosity targets was reduced from five hours to five minutes.
- Usable capacity was increased by several million pounds per year.
- Most significantly, finished product inventory was cut in half.

It should be emphasized that no equipment was relocated. No significant equipment modifications were required. There was some process piping removed, and some valves were locked out so that the cellular boundaries couldn't be crossed inadvertently. Lines were painted on the floor to designate cell assignment, and signs were hung over each process vessel or machine to indicate its cell. Thus, there was some slight cost involved, but miniscule compared to the benefits.

The following list summarizes all the benefits that typically accrue from application of cellular manufacturing concepts to process plants. The application of cellular concepts to the manufacture of industrial fibers, plastic films, automotive paints, and paperlike sheet goods have all resulted in dramatic improvement in the areas listed.

- Shorter, simpler setups
- Less yield loss on restart
- Shorter production campaigns
- Increased throughput
- Reduced WIP

- Reduced finished product inventory
- Shorter manufacturing lead times
- Less dependence on forecast accuracy
- Lower safety stock requirements
- Improved customer service
- Reduced product variation
- Improved material flow
- Improved flow visibility
- Simplified production scheduling
- Easier implementation of pull replenishment systems
- Smaller supermarkets required to support pull

STEPS IN VIRTUAL WORK CELL DESIGN

The step-by-step procedure to follow in designing virtual cells for a process industry plant is summarized as:

1. Update the value stream map.
2. Propose preliminary virtual cell layout.
3. Determine initial product groups.
4. Assign each product group to a virtual cell.
5. Define swing products.
6. Review the design with all stakeholders.
7. Document the plan and the operating rules.
8. Identify each cell visually.
9. Modify scheduling processes.
10. Develop flow-managing processes.

Step 1: Start with the Current State Value Stream Map

Make sure that there is an up-to-date VSM, and that it:

- Indicates how many pieces of equipment are at each step.
- Notes any differences in equipment capability (see Chapter 4).
- Has completed data boxes.

- Shows any equipment-related flow constraints (A1 can't go to B3, for example), including those imposed by process piping or by material handling equipment layout.

Step 2: Determine Preliminary Asset Groups or Virtual Cells

- The goal is to have only one route through each cell, unless a cell has parallel equipment, as was the case with the generic example back in Figure 11.11.
- If capacities of similar pieces of equipment differ significantly, try to group for synchronized throughput through each cell.
- Aim for flow simplicity and flow visibility. Consider geography, but don't let it override equipment capability factors.

Step 3: Determine Preliminary Product Groupings (Group Technology)

- List each product or SKU to be manufactured.
- Calculate takt for each product.
- List any special processing requirements for that product, especially those that would influence the grouping into product families.
- Develop a matrix showing all product requirements that might be relevant to grouping: processing temperatures, additives, viscosity targets, colors, and so on.
- Develop a matrix showing the transition time for each product to go to each other product (similar to the city-to-city mileage chart shown on maps; see Figure 11.16 for an example).
- Group the products into families based on process requirement similarities and simplicity of transitions.

Step 4: Assign Each Product Group to a Manufacturing Cell

- Check the total takt required for each family, and compare it to the effective capacity of each piece of equipment in the cell to ensure that capability is balanced with demand.
- This is an iterative process. The product groupings may suggest modification to the virtual cell arrangement. The reverse is also true: when product groups are assigned to cells, capacity limitations may suggest a product move to a different group.
- Continue the iteration until a reasonable balance is achieved.

Product Changeover Time Matrix (All values in hours)

From/To	F1	F2	F3	F4	J1	J2	J3	J4	R1	R2	R3	R4	R5	R6
F1		2	2	2	7	7	7	7	7	9	10	10	10	10
F2	2		2	2	7	7	7	7	7	9	10	10	10	10
F3	2	2		2	7	7	7	7	7	9	10	10	10	10
F4	2	2	2		7	7	7	7	7	9	10	10	10	10
J1	7	7	7	7		3	3	3	8	9	10	10	11	11
J2	7	7	7	7	3		3	3	8	9	10	10	11	11
J3	7	7	7	7	3	3		3	8	9	10	10	11	11
J4	7	7	7	7	3	3	3		8	9	10	10	11	11
R1	7	7	7	7	8	8	8	8		3	3	5	5	5
R2	9	9	9	9	9	9	9	9	3		3	5	5	5
R3	10	10	10	10	10	10	10	10	3	3		5	5	5
R4	10	10	10	10	10	10	10	10	5	5	5		4	4
R5	10	10	10	10	11	11	11	11	5	5	5	4		4
R6	10	10	10	10	11	11	11	11	5	5	5	4	4	

FIGURE 11.16
Product changeover-time matrix.

Product	Weekly Volume (thousand pounds)		
	Cell F	Cell J	Cell R
F1	500		
F2	*100*	*100*	*100*
F3	50		
F4	50		
J1		450	
J2		150	
J3		50	
J4		50	
R1			130
R2			85
R3			50
R4			45
R5			35
R6			30
R7			25
Total Pounds	700	800	500

Effective Cell Capacity	1000	1000	1000

FIGURE 11.17
Identify a few swing products.

Step 5: Define a Few Swing Products

These are products that can be made conveniently on cells other than their primary cell. This can help with the product-to-cell balancing in Step 4. This situation is shown in Figure 11.17, where product F2 has been made a swing product to better balance takt to capacity. Swing products also help to balance capacity with short-term demand variability.

- Try to minimize the number of swing products, as there can be negative impact on quality and on flow visibility.
- A swing product begun in one cell should follow that cellular flow through the entire process. A specific batch of swing product should not be allowed to cross cell boundaries.

Step 6: Review the Plan

Review the plan with all process owners, and with anyone who would have relevant input. This would normally include:

- Operators, who normally perform the changeover tasks
- Mechanics, who may also be involved in equipment setup tasks
- First-line supervision
- Process engineers and product technologists, who may have additional insight into product requirements and product-to-product transition considerations
- Production schedulers

If possible, all of these stakeholders should be involved in the complete cell design process. A kaizen event (see Chapter 9) would be an excellent way to do this.

Step 7: Document Virtual Cell Arrangements, Flow Patterns, Product Lineups, and Operating Rules

The future state VSM should reflect a cellular arrangement, as shown in Figure 11.18 for our sheet forming process.

Step 8: Mark Each Cell Visually

Hang a sign over each piece of equipment to designate the cell to which it belongs. Paint flow lines on the floor. Use color coding and other visual techniques to identify the components of each cell.

Step 9: Modify Scheduling Processes Accordingly

This is a revolutionary change for most production schedulers. Make sure that they understand the new flow patterns and rules, and what is to be gained by trading flexibility for greater structure and flow discipline.

Step 10: Ensure that Appropriate Managing Processes Are in Place

Because of the traditional value for flexibility, it takes discipline to maintain the virtual cell plan. This is easier to do in parts machining and assembly processes, where equipment has been relocated into physical cells; the new physical arrangement makes it very difficult to revert to old flow patterns. It is far easier to deviate from *virtual* cellular flow, and tempting to do so when problems arise. Contingency planning, predetermining how

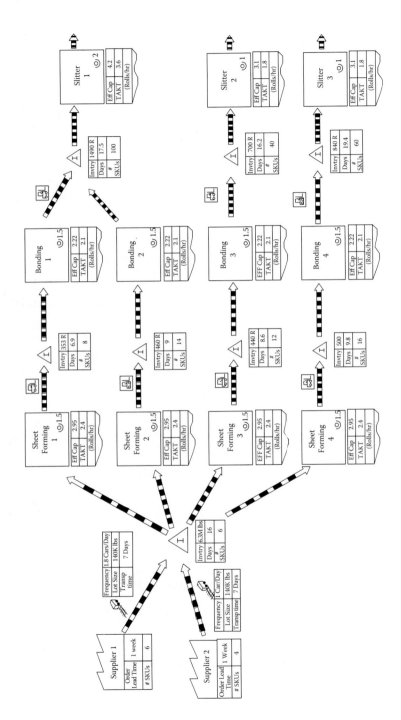

FIGURE 11.18

Document the flow—the future state VSM.

to deal with unusual events, can help mitigate the perceived need to break the plan.

Where the equipment configuration lends itself to cellular manufacturing, it should be designed and implemented before pull replenishment systems are considered. It should be apparent that pull is far easier to put in place with the small number of flow paths that a cellular arrangement requires than with the large number of flow paths possible with a functional layout.

SUMMARY

Viewed in terms of Ohno's seven wastes described in Chapter 3, flow based on a traditional functional layout can cause most of the seven:

- **Overproduction:** A functional layout generally leads to longer campaigns and, thus, more production than is currently required.
- **Inventory:** Long campaigns create excess inventory.
- **Transportation:** The inventory must be stored somewhere, and in process industry plants this can be in a fairly remote area.
- **Waiting:** Product changes can be quite long on equipment that must run the full range of products, so this causes more time spent waiting.
- **Defects:** Product changes from one major family to another can result in significant yield loss on start-up.
- **Processing:** These yield losses don't disappear on their own; someone must load them into some type of container for transportation to a waste disposal area. The task of moving them there is another cause of transportation waste. In some cases the yield losses can be reworked to become first grade, or be blended in small percentages with good product, both of which are examples of wasteful processing.

Cellular manufacturing can reduce or eliminate all these wastes. It also addresses the eighth waste: If the operators, mechanics, and schedulers who operate the process on a daily basis are included on the cell design team, it puts their intelligence, experience, and creativity to good use.

The final thought, and indeed the main message of this chapter, is that anyone in the process industries who has dismissed cellular manufacturing as impractical or irrelevant has simply not looked at cellular manufacturing from the right perspective. The key is to think *flow,* not *function.*

12

Product Wheels: Production Scheduling, Production Sequencing, Production Leveling

Most manufacturing operations will run much more efficiently if production is done at a uniform, level rate. This tends to maximize equipment productivity and labor productivity, and smooth out the requirement for raw materials and support facilities. Unfortunately, for most plants, demand is not uniform at all. Variation in demand can be due to several factors: the normal, random variation that will almost always be present; seasonality; and longer-term trends due to new product acceptance or to sales decline of mature products.

Any variation in production rate can create waste. From Ohno: "On a production line, fluctuations in product flow increase waste. This is because the equipment, workers, inventory, and other elements required for production must always be prepared for peak production." Variation in production rate will also cause variation in raw material consumption and, thus, push some degree of the variation and waste back to suppliers. Further, starting and stopping production equipment often degrades the quality of the material being produced, creating more waste.

As you will see in this chapter, lean addresses these wastes by driving toward level production and smooth continuous flow.

SOLUTIONS IN ASSEMBLY PROCESSES

The goal of lean is continuous flow or as close to that as possible, to produce at a constant rate, to eliminate the waste of waiting and the waste of over-production. Recognizing that demand can be quite variable, a means to produce to average demand must be found. Understanding the true meaning of takt is helpful in this regard. Takt, the rate of customer demand, is not an instantaneous measure, but demand averaged over some period of time. Takt is often calculated on a daily basis, which will smooth out the short-term variation and, thus, buffer the manufacturing operation from short-term fluctuations in customer demand. If demand has longer-term variability, takt should be calculated accordingly: Takt should represent the average demand over reasonable time periods, and should be recalculated periodically to accommodate longer-term variability.

Heijunka is the lean technique for smoothing production to meet average customer demand. As described by Jeffrey Liker in *The Toyota Way*, "Heijunka is the leveling of production by both volume and product mix. It does not build products according to the actual flow of customer orders, which can swing up and down wildly, but takes the total volume of orders in a period and levels them out so the same amount and mix are being made each day." Leveling the *mix* is important to balance the use of people, equipment, and parts. If different parts to be produced on a line require different operations and, therefore, have different labor content, the utilization of labor can be smoothed out by producing a small number of each rather than producing in large lots. Similarly, if different parts require different machine time and different materials, these will all be smoothed out by producing in small lots. Figure 12.1 illustrates this practice, called mixed-model scheduling; if our product mix is 60 percent part A, 20 percent part B, and 20 percent part C, a mixed model production sequence would suggest that production be in the order of two As, one B, another A, a C, and then repeat the sequence.

Leveling production volume can be done by collecting all orders over the appropriate time span and scheduling them to be produced at an even rate, and in the mixed sequence. A heijunka box (Figure 12.2) is the normal way this is accomplished in lean assembly processes. As requirements

FIGURE 12.1
Mixed model production sequence.

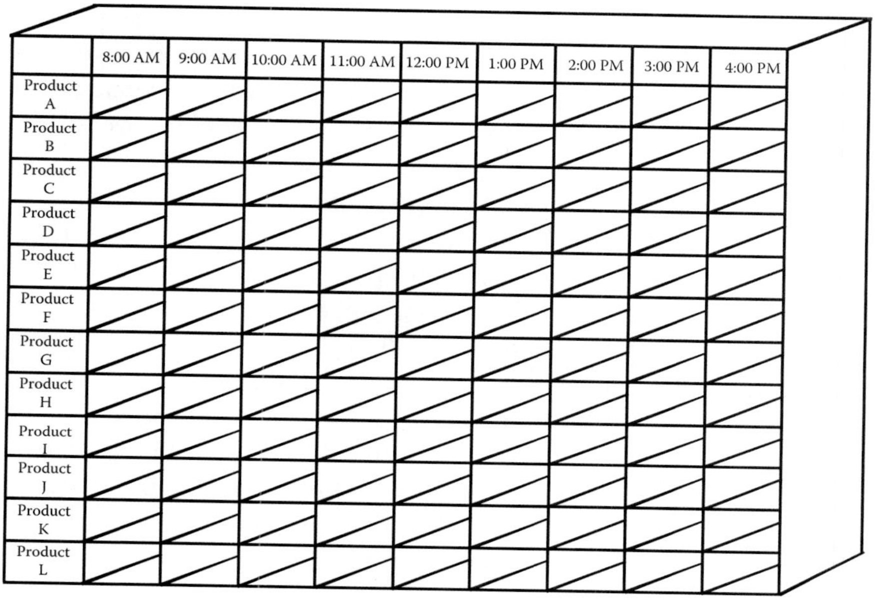

FIGURE 12.2
A *heijunka* box.

for product A are received, cards representing specific amounts of A are placed in the first horizontal row of the box, with cards for the defined mixed-model quantity being placed at 8 a.m., at 9 a.m., and so on. The cards are removed from the box at the indicated times to allow production of those parts. Thus, the goal of leveling volume is accomplished by the time structure of the vertical columns, and the goal of leveling mix is accomplished by the horizontal part-type structure.

Like most lean concepts and tools, the heijunka box originated with Toyota, having evolved from a combination of their maintenance scheduling process and their kanban cards.

PROCESS INDUSTRY CHALLENGES

Conceptually, everything just described can work in process industry operations, and can offer the same advantages. However, there are some challenges that require a slightly different perspective when applying them.

- Demand variation is typically over much longer periods of time, weeks, months, or over a full year for some food products, beverages, and crop fertilizers and pesticides. Because many process operations run near full capacity, production leveling must be done over longer periods to avoid shortages during peak seasons.
- The difficult and costly product changeovers often found in process equipment, and the frequently accompanying yield losses on re-start, make mixed-model scheduling impractical within short time periods. In the typical sheet goods process, it would be very costly to make a single nine-foot-wide roll, then a ten-foot-wide roll, then another nine-foot-wide roll, and so on. Continuous chemical reactions cannot practically make one tank of A, then one tank of B, and then back to one tank of A. A mixed model scheduling strategy must, therefore, accommodate longer campaigns for some products.
- Another avenue sometimes taken in assembly operations is to use overtime to manage surges in demand, so that production per hour can remain level. Because many process plants are already running a 24-hour-per-day operation, this is usually not an option.

━━━━━━━━━━

A PROCESS INDUSTRY SOLUTION: THE PRODUCT WHEEL CONCEPT

A technique that has evolved in some process industry operations addresses these challenges in a reasonably straightforward manner, while accomplishing the dual objectives of production leveling and mixed-model scheduling to the optimum practical extent. It brings with it a third advantage, that of optimizing the sequence of the mixed-model schedule. Changeover time, difficulty, and cost are often dependent on the sequence in which products are produced, so sequencing is a critical feature.

The concept is called product wheel, which is a visual metaphor for a structured, regularly repeating sequence of the production of all the materials to be made on a specific piece of equipment, within a reaction vessel, or within a process system. One cycle of the product wheel could be considered to be one column of a heijunka box; it is the time period over which the mixed-model quantity of each product is made.

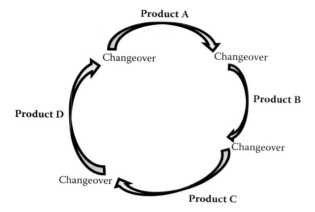

FIGURE 12.3
The product wheel concept.

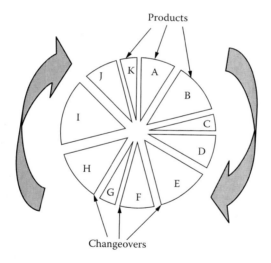

FIGURE 12.4
Detailed product wheel concept.

Figure 12.3 illustrates the concept for a wheel with four products, A, B, C, and D. Each is made in the appropriate quantity, determined by takt for that product, and then the cycle is repeated. Figure 12.4 shows a wheel in somewhat more detail, this time for a wheel with eleven products, A through K. The portion of the cycle where a single product is being made is called a spoke, so this wheel has eleven spokes. As diagrammed, the spokes have differing lengths, reflecting the differing takt value for each product.

In concert with lean principles, the goal is to design the wheel to have as short a total cycle time ("wheel time") as is feasible. That of course implies that each spoke will be as short as possible, in keeping with one-piece-flow ideals. Product wheel design approaches wheel time determination from two independent perspectives:

1. Given the portion of the total available time that is needed for production, how much time remains that could be used for changeovers? If all of that is used for changeovers, how fast can the wheel spin?
2. Considering that changeovers have a cost penalty, what is the optimum economic trade-off between changeover cost (greater with faster wheels) and inventory carrying cost (greater with longer wheels)?

Thus, the mixed-model aspect of heijunka is accomplished, but tempered with the economic realities of process industry changeovers. The volume leveling aspect is also satisfied by designing each spoke to produce to the takt for that product, leveled over the time period covered by the wheel time.

Even if production leveling weren't an objective, product wheels would still be beneficial in providing a methodology to optimize the sequence of products to be made on an asset, and in providing a process to optimize the length of the lot size or the campaign length for each product, balancing changeover cost with inventory carrying cost.

Product wheels should, therefore, be considered for any piece of equipment, vessel, or process system where changeovers have costs associated with them, and especially where these costs are dependent on the specific sequence followed.

Product wheels can be employed in a make-to-stock (MTS), make-to-order (MTO), or finish-to-order (FTO) environment. In fact, MTO and MTS products can be produced on the same wheel, as will be seen in the example later in this chapter.

A product wheel can be operated in a push or in a pull mode. The specific differences, primarily how demand signals are loaded onto the wheel, are discussed in Chapter 14; the examples in this chapter generally describe a pull operation. That is, the wheel will be *designed* based on average historical demand or on forecast demand, but what is *actually produced* on any spoke is just enough to replenish what has been consumed from the downstream inventory.

Product wheels can be employed at several steps in the total manufacturing process. In a typical parts assembly manufacturing operation, it is recommended that a pacemaker be determined for the entire process, that it be the single scheduling point for the entire operation, and that heijunka be used to schedule the pacemaker. In process industry production lines, the concept of a single pacemaker is not always appropriate; there is often the need to schedule some steps independently of the others. In the sheet goods process introduced in Chapter 2, the factors that would determine an optimum schedule on the forming machines may be different from the factors influencing scheduling on the bonders. The grouping of products into families may yield different combinations at each step, with forming products grouped by polymer type or sheet width and bonder products grouped by bonding temperature. Thus, it may be appropriate to have product wheels at forming and bonding, running independently of each other, so both forming and bonding would be pacemakers. In that case, there would likely be an independent product wheel on each forming machine, each running a different cycle with perhaps different wheel times, each optimized for the products made on that wheel. Similarly, each bonder may be running an independent wheel, as mapped in Figure 12.5. This figure represents another step in the evolution of the future state VSM, after product wheels have been applied to the cellular manufacturing concepts described in Chapter 11.

In an MTS environment, there must be inventory downstream of any process step using product wheel scheduling, so that customer orders for a specific product can be satisfied while other products are being made on the wheel (that is, on the asset being managed by the wheel). This inventory, which could be in a downstream WIP location or in the warehouse as finished product, allows the operation to manage variability, by producing to takt, to the level demand averaged over the wheel time.

Just as the heijunka box provides a way to level production volume and mix over time, product wheels provide a way to level volume and mix over the cycle of the wheel. Determining the optimum wheel design parameters is now described.

PRODUCT WHEEL DESIGN

The best way for you to understand product wheels is to walk through the wheel design process.

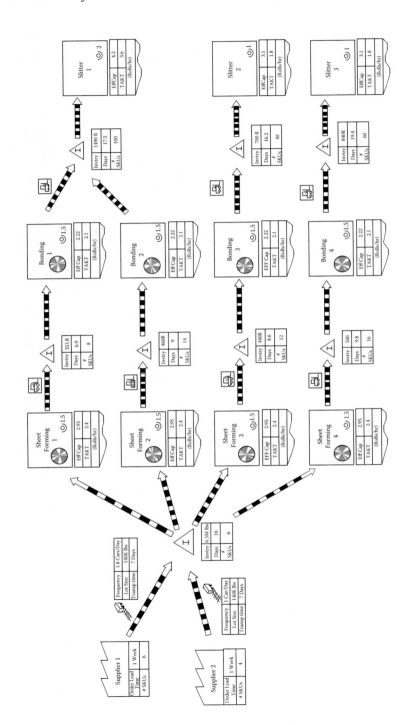

FIGURE 12.5

Future state VSM showing assets to be scheduled by product wheels.

Step 1: Determine Which Process Steps Should Be Scheduled by Product Wheels

Any step in the process, any piece of equipment or process system, which has appreciable changeover times or losses, should be examined as a candidate for a product wheel. "Appreciable" in this context means any changeover long enough or experiencing enough material loss that it affects the scheduling of the step. In these situations, the product wheel methodology will help to set the optimum campaign lengths for all products made on that asset. If the time, difficulty, or material loss is sequence dependent, it is even more likely that product wheels will be beneficial, because of the sequence-determining part of the methodology.

Referring to the VSM of our sheet forming process in Chapter 4, the product changeover time for the sheet forming step was listed on the map as one hour. The forming machine has automatic roll changing capability on the windup end, so that as a roll has completed the winding operation, it rotates out of the sheet path and an empty core is positioned to receive the sheet. Thus, if subsequent rolls are of the same product type, the transfer is instantaneous. If, however the product type changes, time will be lost, and the one hour shown is an average of all types of changes. If the width of the sheet is changed, say from nine feet to ten feet, that is an easier change and takes much less than an hour. If the basis weight is to be changed, by altering the rate at which material flows onto the forming belt, that takes longer. If the raw material type is changed, transfer lines must be emptied and cleaned, resulting in the longest changeover. Thus, to minimize total changeover time there is an optimum sequence, so the forming machines are logical candidates for product wheels.

Likewise, the bonders are candidates for product wheels. The various products must be bonded at different temperatures, and it takes significant time to change bonder temperature. The surface of the bonder roll is precisely machined to be flat within very small tolerances, and heating or cooling it rapidly could warp the surface. For that reason it is best to sequence products so that the temperature change at each product change is small.

The slitters are a different story. The changeovers are quick, averaging five minutes or less, with no appreciable yield loss, so we have elected not to schedule them using the product wheel methodology.

So the VSM shown in Figure 12.5, which reflects the cellular manufacturing improvements described in the last chapter, now shows a product wheel symbol on each forming machine and each bonder, to indicate that our future state will incorporate them being on wheels. The wheels on each of these pieces of equipment will run independently of each other, and may run different wheel cycle times.

Step 2: Analyze Product Demand Variability

The next step in product wheel design is to examine the average demand and the demand variability of each product to be made on a wheel, to decide whether that product is best made to order, finished to order, or made to stock. Figure 12.6 shows a Pareto chart of average demand for a typical product lineup, and illustrates the common trait that a small portion of the total product portfolio has most of the demand volume. (Pareto predicts that 20 percent of the products have 80 percent of the demand.)

Once each product has been analyzed for average demand and demand variability, the results can be placed on a decision matrix (Figure 12.7), which can guide decisions on MTO versus MTS. Products in the lower right quadrant, with high demand and low variation, are the best candidates for MTS, because there is little risk in carrying inventory of those items; they are likely to sell soon. Products in the upper left quadrant are best made to order: Because demand is low and quite variable, if they were made to stock, they could sit in inventory for long periods of time.

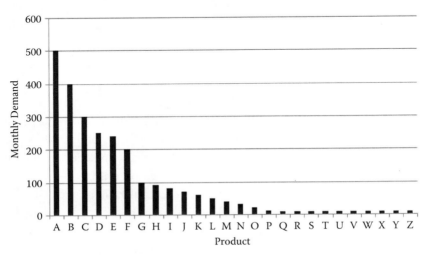

FIGURE 12.6
Pareto chart of demand by product.

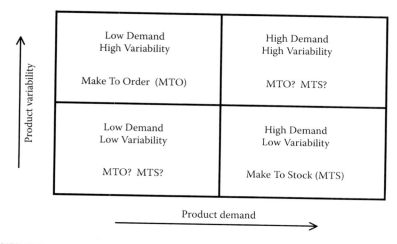

FIGURE 12.7
MTO/MTS decision matrix.

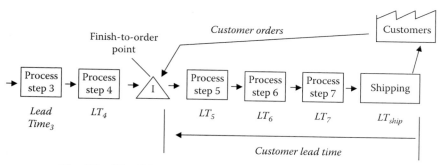

$LT_5 + LT_6 + LT_7 + LT_{Ship}$ must be less than *Customer lead time*

FIGURE 12.8
Lead time requirements for FTO.

One requirement for making to order or finishing to order is that the total manufacturing lead time and shipping time, from the MTO or FTO point to arrival at the customer, must be less than the delivery lead time committed to the customer, as shown in Figure 12.8. If this condition is met, FTO is a realistic alternative. If not, and FTO appears beneficial, work should be done to reduce lead time enough to make FTO practical. (FTO is covered in more detail in the next chapter.)

Let's take a closer look at how these tools apply to sheet Forming Machine 1. The product grouping done as part of cell design has resulted

Product	Weekly Demand D (Rolls)	σ_D (Rolls)	Coefficient of Variation $CV = \sigma_D/D$
A	150	30	0.20
B	120	20	0.17
C	75	24	0.32
D	15	6	0.40
E	14	6	0.43
F	12	5	0.42
G	8	6	0.75
H	6	3.6	0.60

FIGURE 12.9
Demand variability analysis.

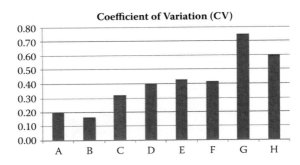

FIGURE 12.10
Demand variability for products made on Forming 1.

in eight products (named A through H) being assigned to Forming Machine 1. Figure 12.9 shows the average weekly demand and demand variation for each product. It also lists the coefficient of variation (CV), a measure of the relative variability:

$$CV = \frac{\sigma_D}{D_{AVG}}$$

Figure 12.10 shows CV plotted as a bar chart; it can be seen that products A and B have stable demand patterns, while C, D, E, and F are somewhat variable, and G and H are much more variable. Putting this and the demand data on the decision matrix results in Figure 12.11, which suggests that A, B, and C be made to stock, G and H be made to order, and that D,

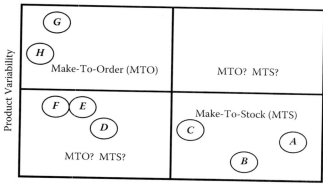

FIGURE 12.11
Decision matrix for products made on Forming 1.

E, and F might go either way. The resulting decision is that A through F will be made to stock, and that G and H will be MTO. The sales policy for G and H will be set so that the delivery lead time quoted to customers will be longer than the manufacturing/shipping lead time; this is often the policy for low volume, sporadic products.

MTO is being suggested in this case so that the inventory for a small number of low volume, high variability products can be eliminated. Chapter 13 discusses the use of MTO and FTO as a more general strategy to manage the production of entire product portfolios with very high degrees of differentiation or customization.

Step 3: Determine the Optimum Sequence

The next step in wheel design is to determine the optimum sequence in which the various products should be made. In some cases this is obvious. With Forming Machine 1, we know that we should make all the widths requiring a specific raw material type and the same basis weight, then change to the products with a different basis weight but with the same raw material feed, and finally change the raw material feed. So for optimum sequencing the products should be grouped first by feed material type, then within types by basis weight, and finally within basis weights by width.

In many cases the optimum sequence is not so obvious; the parameters affecting changeover difficulty can be more complex and are often interrelated. In those cases it is helpful to develop a from–to changeover matrix like the one shown as Figure 11.16.

Step 4: Calculate Shortest Wheel Time Possible (Available Time Model)

Once the sequence of products on the wheel has been established, the next step is to decide how long a single cycle of the wheel should be. Two methods are used to give guidance on optimizing wheel time:

1. Available time model
2. Economic order quantity (EOQ)

These will generally give different answers, and neither should be considered the best answer; they each give useful perspectives that should influence the wheel timing decision.

The available time model considers some period of time, perhaps a week. It then computes the amount of time required to produce the full customer demand for that period and subtracts that from the total available time. The difference is the amount of time that could be available for changeovers. The total changeover time for one cycle of the wheel is calculated by adding up all the individual product changeovers. That sum is then divided into the total time available for changeovers to indicate how many cycles could be completed per week. The wheel time can be the number of hours per week divided by the number of cycles.

$$\text{Wheel cycles per period} = \frac{\text{Total available time} - \text{Total production time}}{\sum \text{Changeover times per cycle}}$$

$$\text{Wheel time} = \frac{\text{Total available time}}{\text{\# of Wheel cycles per period}}$$

With a takt of 2.4 rolls/hour, Forming Machine 1 should produce 400 rolls per week. At the effective capacity of 2.95 rolls/hour, that will require 136 hours of actual production, leaving 32 hours (168 – 136) for changeovers. With the virtual cell lineup in place, Forming Machine 1 produces eight product types, requiring an average of one hour each, so a complete set of changeovers consumes eight hours.

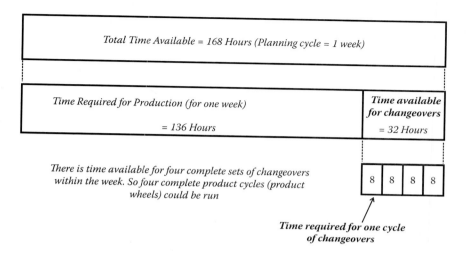

FIGURE 12.12
Available time analysis for Forming 1.

This is shown graphically in Figure 12.12 for sheet Forming Machine 1. As 32 hours are available for changeovers each week, and one complete set takes eight hours, we could run four wheel cycles per week, of 42 hours each. Thus, the wheel time would be 42 hours. This represents the shortest wheel time possible under the stated conditions; it is not to suggest that it is the best wheel time, just the minimum possible.

This assumes that every product is made on every wheel cycle, which might not be the case, but it is a reasonable assumption to start from.

Step 5: Estimate Economic Optimum Wheel Time (the EOQ Model)

The available time calculation only provides the shortest wheel possible under the current conditions, but not necessarily the best. The EOQ model takes into account the economics involved, and thus gives another perspective.

The EOQ is a classic way to calculate lot size or campaign size. It approximates the best balance between total changeover cost and inventory carrying cost. Larger campaigns, with their longer product wheels, require more inventory to supply customers or downstream operations while specific products are not being made, and smaller campaigns require more changeovers.

The specific equation for the quantity that results in the lowest total cost is:

$$EOQ = \sqrt{\frac{2 \times COC \times D}{V \times r \left(1 - \dfrac{D}{PR}\right)}}$$

where COC = changeover cost, D = demand per time period, V = unit cost of the material, r = % carrying cost of inventory per time period, PR = production rate in units per time period.

(Note that the time period must be the same for all factors. If demand is in rolls per week, r must be the annual percentage divided by 52.)

A simpler equation for EOQ is sometimes presented:

$$EOQ = \sqrt{\frac{2 \times COC \times D}{V \times r}}$$

The equation being recommended here accounts for the fact that some of the material produced is being consumed during its spoke, so that the inventory at the end of the spoke is slightly less than the sum of the amount produced and the starting inventory. This equation is sometimes called economic production quantity, or EPQ.

Figure 12.13 shows the EOQ concept graphically. The inventory cost can be seen to rise with wheel length, the changeover cost declines due to the decreasing number of changeovers, and total cost drops and then rises. The total cost has its minimum at the wheel time where the two cost components intersect.

It must be noted that the EOQ calculation is only an approximation, that it takes into account the factors that typically are most significant, but ignores some factors usually having lower impact. The inventory included, for example, is only the cycle stock required, that is, inventory to satisfy average demand between production spokes for a material. It ignores safety-stock, inventory needed to protect against normal random variation in supply and demand.

An interesting property of the EOQ curve is that total cost is flat in the region of the minimum. The consequence of this is that selecting a wheel time somewhat greater or less than the optimum will be nearly as good. Selecting a wheel length 20 percent lower than the optimum will increase

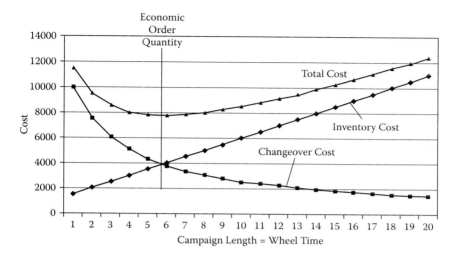

FIGURE 12.13
EOQ concept.

total cost by only 3 percent; selecting a 30 percent longer wheel time also increases total cost by only 3 percent.

These two factors, the flatness of the curve and that fact that it is only an approximation, reinforce that the EOQ result should not be considered as the best answer, but only a guide. Practical considerations must also be considered. Many operations function more smoothly if the wheel time is some integer number of days or weeks. If the wheel time is fixed at one week, that can make it easier to follow. Everyone involved, schedulers, planners, operators, maintenance, and test labs, can get into a regular pattern. If there is excess capacity, the wheel will end before the next cycle must begin, and if this occurs at regular, repeatable times, preventative maintenance can be scheduled in advance for the idle periods.

Figure 12.14 shows the results of the EOQ calculation for paper Forming Machine 1, the EOQ values and the frequency (wheel time) suggested for each of the eight products.

Step 6: Determine the Wheel Time (Making the Choice)

With the available-time calculation suggesting a wheel time of approximately two days (forty-two hours), and EOQ suggesting times ranging from about seven days to thirty days, a choice must be made. Although a

Product	Weekly Demand D (Rolls)	Cost per Roll	Change Over Cost	Inventory Carrying Cost	EOQ (ROLLS)	Optimum Frequency (Days)
A	150	$1,800	$450	25%	158.0	7.37
B	120	$1,800	$450	25%	133.5	7.79
C	75	$2,000	$500	25%	98.0	9.14
D	15	$2,000	$500	25%	40.3	18.79
E	14	$2,000	$500	25%	38.8	19.42
F	12	$2,400	$600	25%	35.9	20.92
G	8	$2,400	$600	25%	29.1	25.50
H	6	$2,400	$600	25%	25.2	29.36

FIGURE 12.14
EOQ analysis for Forming Machine 1.

two-day wheel is possible, and would be the preferred choice from a pure lean perspective in that it makes the smallest lot sizes, changing product type that frequently would heavily utilize maintenance mechanics and test lab facilities, well beyond current capacity.

The EOQ results suggest that a seven-day wheel be selected, and products A, B, and C be made every cycle. (The basic wheel time is usually set by the high volume products.) Products D, E, and F would fit well being made every twenty-one days, every third wheel cycle. Note that none of the calculations resulted in exactly seven or twenty-one days, but the flatness of the EOQ curve means that these are almost equally good choices. The demand variability analysis done earlier had suggested that G and H be made to order; doing so will still require that spokes be available on the wheel for periods when orders for those products are received. Figure 12.15 shows the planned wheel and how each product would be placed on various cycles.

Figure 12.16 is a more visual representation of the wheel design. Each wheel is seven days; lower volume products are made every third wheel cycle and are distributed to cycles so that overall production requirements are relatively equal from cycle to cycle. Empty spokes have been allocated to G and H in case orders are received. Scheduling of specific products has been sequenced in accordance with the analysis discussed in Step 3. The totals at the bottom of Figure 12.15 validate that production requirements are balanced from cycle to cycle.

Product	Weekly Demand D (Rolls)	EOQ	Optimum Frequency (Days)	Recomd'n (Days)	Cycle 1 (7 Days)	Cycle 2 (7 Days)	Cycle 3 (7 Days)
A	150	158.0	7.37	7	150	150	150
B	120	133.5	7.79	7	120	120	120
C	75	98.0	9.14	7	75	75	75
D	15	40.3	18.79	21	45		
E	14	38.8	19.42	21		45	
F	12	35.9	20.92	21			36
G	8	29.1	25.50	MTO			25
H	6	25.2	29.36	MTO		20	
Totals					390	410	406

FIGURE 12.15

Wheel time determination for Forming Machine 1.

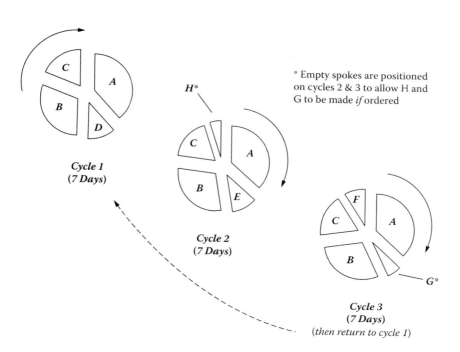

* Empty spokes are positioned on cycles 2 & 3 to allow H and G to be made *if* ordered

Cycle 1
(7 Days)

Cycle 2
(7 Days)

Cycle 3
(7 Days)
(*then return to cycle 1*)

FIGURE 12.16

Product wheel design for Forming Machine 1.

Step 7: Calculate Inventory Requirements

With MTO products, no inventory is needed to support the product wheel. As product orders are received, they are loaded onto the wheel schedule (equivalent to placing cards into the appropriate slot in a heijunka box). After they are produced on the wheel, they flow through the downstream steps, through to packaging and truck loading. They may stop temporarily at a buffer inventory, but for reasons other than requirements imposed by the wheel schedule.

For MTS products, inventory will be required, proportional to wheel time. For these products, there must be inventory downstream, either as WIP or as finished product, to satisfy needs for that material during the period when other products are being produced on the wheel. Figure 12.17 tracks the inventory for a single product made on a wheel. An amount, P1, of that product is produced at the beginning of Wheel Cycle 1, and flows through downstream process steps into the inventory shown in Figure 12.17. Demand for that product during the remainder of the wheel will consume portions of that inventory until the next production, P2, arrives at the inventory. This rising and falling of inventory in a sawtooth pattern repeats for every cycle of the wheel.

If demand were always exactly equal to the average, only cycle stock, the average demand during one wheel cycle, would be needed. However, demand will typically vary in a random fashion, sometimes greater than average and sometimes less. To avoid stockouts during periods of higher

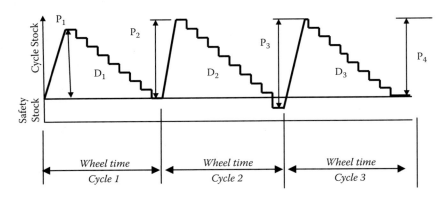

D = Demand During the Wheel Cycle
P = Production to Begin a Wheel Cycle

FIGURE 12.17
Inventory required to support the wheel (cycle stock and safety stock).

demand, safety stock is usually carried. Safety stock can also provide protection from stockout in cases where the wheel, due to production problems, doesn't return to this spoke at the scheduled time. The demand, D2, in Period 2, is higher than average and the safety stock provides material to supply downstream needs and customers until production P3 arrives.

(Some authors use the term "safety stock" for the inventory carried to protect against production problems or lead time variation, and "buffer stock" for that to protect against demand variation. Others use the terms in the opposite sense, while still others use "safety stock" to describe the combination. I follow the last convention.)

It should be apparent that shorter wheels require less cycle stock, and longer wheel times more cycle stock. Safety stock is also somewhat proportional to wheel time; shorter wheels mean that the variability causing the need for safety stock will be less (this is true if the variables follow a normal distribution).

The calculation of cycle stock is straightforward. It is the average demand for that product for one wheel cycle time, times the cycle frequency if the product is not made every cycle. If a product is made every third cycle, for instance, the cycle stock must support demand over three wheel cycles. Thus, the cycle stock for product D is forty-five rolls, resulting from a weekly demand of fifteen rolls and a plan to make it every third cycle.

Cycle Stock = Average Demand per unit Time × Cycle Frequency

The calculation of safety stock is somewhat more complex, and is covered in detail in the chapter on supermarket design.

The total inventory required to support the product wheel should be calculated. This includes the cycle stock and safety stock for all MTS products made on the wheel. Figure 12.18 tabulates total inventory for Forming Machine 1. It should be noted that the total inventory at the completion of a production spoke for a product will not quite equal cycle stock plus safety stock. This is because some of the material produced was consumed during its spoke. This can be significant for products occupying a large portion of the wheel: Product A occupies 37 percent of the full wheel cycle, so 37 percent of its production is consumed, on average, before the end of the A spoke. Thus, although its cycle stock is 150 rolls, 56 of those get consumed during the production, so the inventory at the end of that spoke will be 94, plus the safety stock of 50 rolls, for a total of 144 rolls.

224 • *Lean for the Process Industries: Dealing With Complexity*

Product	Weekly Demand D (Rolls)	Cycle 1 (7 Days)	Cycle 2 (7 Days)	Cycle 3 (7 Days)	Cycle Stock (Rolls)	Safety Stock (Rolls)	Maximum Inventory (Rolls)	Average Inventory Required (Rolls)
A	150	150	150	150	150	50	144	97
B	120	120	120	120	120	33	117	75
C	75	75	75	75	75	40	101	70
D	15	45			45	18	62	40
E	14		42		42	18	60	38
F	12			36	36	15	51	32
G	8			25	0	0	0	0
H	6		20		0	0	0	0
Totals	400	390	407	406	468	174	534	353

FIGURE 12.18
Maximum and average inventory required to support the wheel.

Step 8: Fine-Tune the Design

The first cut at a wheel design is now complete. However, things sometimes don't work out that neatly in the real world, and adjustments are often required. For example, the inventory calculated in Step 7 may require more space than is available. The inventory could be stored at a more remote location, but the transportation would be waste, the very thing we're trying to eliminate. Or, the carrying cost of the inventory could be more than the business thinks is affordable. In some cases the EOQ model might suggest a short wheel, but the available-time calculation indicates that there is not enough free capacity to spin the wheel that fast.

For a variety of reasons the wheel may not be feasible as first designed, so one should be prepared to recycle through the steps; usually two or three iterations result in a workable wheel design. The fact that the EOQ curve is both approximate and flat allows quite a bit of latitude.

Although in this example, the support facilities wouldn't permit the short wheel time suggested by the available-time model, so that we based wheel time on EOQ, it is often enlightening to work the design to completion on the available-time basis to see if it is feasible considering all manufacturing factors and constraints. Designing the wheel on that basis will always result in the shortest wheel, which satisfies the lean criteria for small lot sizes made frequently. On the other hand, it must be recognized that it may increase the number of changeovers and, thus, the associated waste.

Step 9: Revise the Current Scheduling Process

All formal scheduling processes must be examined and modified as appropriate to accommodate product wheel scheduling. Production planners and schedulers should participate in the product wheel design because they have perspectives that can be valuable to the design process. For successful operation, the production schedulers must understand the reasons for the product wheels and the details of the design.

Step 10: Create a Visual Display

If visual scheduling tools are not currently being used on the plant floor, they should be added, both as a valuable tool in product wheel management and an important step forward in the workforce engagement process. An example of a visual scheduling board—a takt board—was shown for one of the forming machines in Figure 8.2, in the discussion on the role of takt boards as a component of visual management.

The visual displays should provide the same functionality that a heijunka box does. It should show the next cycle of the product wheel, which products are to be made on this cycle and in what order, and the quantity of each. It should display how much time will be left at the end of the current cycle before the next wheel cycle must begin, so that operators and mechanics know how much time will be available for preventative maintenance tasks. It should have an area where operators can record performance to the plan as the wheel progresses, so that all in the area know the current status and so that corrective action can be taken if problems are occurring.

Thus, the visual display boards act as both the heijunka box and as andon (an indication of line status) boards.

The complete product wheel design process can be summarized as:

1. Decide which assets would benefit from product wheels.
2. Analyze product demand variability (consider MTO and FTO).
3. Determine the optimum production sequence.
4. Calculate the shortest wheel time based on time available for changeovers.
5. Estimate the economic optimum wheel time based on EOQ model.

6. Determine the basic wheel time; determine which products get made on every cycle and the frequency for others; balance low-volume products assigned to each product wheel cycle; provide empty spokes for MTO and FTO products.
7. Calculate inventory levels to support the wheel (cycle stock and safety stock).
8. Repeat Steps 3 through 7 to fine-tune the design.
9. Revise all scheduling processes, as appropriate.
10. Create a visual display (heijunka) to manage the leveled production.

BENEFITS OF PRODUCT WHEELS

Product wheels tend to level production as a natural behavior. Because cycle stock is based on customer takt over the period of one or more wheel cycles, production naturally flows to the takt rate. The product wheel design methodology tends to drive mixed model scheduling to smallest practical campaign size. Although the circular wheel metaphor differs from the orthogonal heijunka box structure, they are similar in that an individual wheel cycle can be viewed as one column from the box.

Product wheel design forces a structured analysis of various kinds of changeover to optimize the sequence of production. People sometimes assume that they are following an optimized sequence, but a data-based analysis often reveals that improvements can be made.

Product wheels add predictability to high-variety operations, that is, to operations with a high degree of product differentiation. Everyone associated with an operation scheduled by product wheels knows what is going to be produced and at what time. Any special tools, materials, or personnel required for changeovers can be scheduled in advance. If the start-up of the next material places additional loading on the test lab, this can be planned for.

The design methodology allows a clear understanding of trade-offs between changeover costs and transition losses versus inventory costs. The benefits of further SMED activities can be quantified. Justification for capital improvements that would improve flexibility can be determined.

A fixed, repeatable schedule allows cycle stock requirements to be explicitly known. A fixed span between production cycles of a given material fixes the time increment for standard deviation computation, so that safety stock requirements can be calculated.

The following list summarizes these benefits. Product wheels:

- Tend to level production as a natural behavior.
- Optimize production sequence.
- Add structure and predictability to high-variety operations.
- Provide a basis for informed decisions about production sequence and campaign length.
- Provide a basis for informed decisions about MTO and FTO for appropriate products.
- Optimize transition cost versus inventory carrying cost.
- Provide a structured basis for determining cycle stock requirements.
- Provide a structure basis for calculating safety stock requirements.
- Quantify the benefits available for further SMED activity.

SOME ADDITIONAL POINTS

It should be clear that if an operation has opportunities for cellular arrangements (virtual cells but not necessarily physical cells), they should be designed before product wheels are designed. Product wheel design is far easier if the product grouping phase of cell design has been completed, so that each wheel has fewer products to manage.

Completed SMED activities will also benefit product wheel design. Reducing the changeover time will move the balance point on the EOQ curve toward smaller campaign sizes. It will also reduce the denominator in the available-time calculation, again pointing toward smaller campaigns and faster wheels.

It is highly recommended that during execution of the wheel, production not always match the cycle stock exactly, but be varied based on consumption since the last cycle. The practice should be that at the start of each spoke, enough production is scheduled to bring inventory for that product back to the cycle stock plus safety-stock value. In that mode, sometimes production will be more than cycle stock, sometimes less, and averaging out to the cycle stock value over time. This is in accordance with the pull concepts discussed in Chapter 14.

SUMMARY

Because few manufacturing operations have enough capacity to meet instantaneous demand during peak periods, some form of production leveling is desirable. Variation in demand leads to waste, while production leveling reduces that waste, optimizes equipment and labor utilization, and smoothes out demand for raw materials.

Lean practitioners in parts assembly have traditionally employed heijunka, in the form of boxes or scheduling boards, to level production.

Product wheels provide the leveling function in many process operations. They visualize the production of all product types to be made on an asset as a circular wheel, where each spoke is a different product. Product wheel design starts with making sure that all products are made in a fixed sequence, designed to minimize changeover time and losses. The wheel methodology provides tools to decide how long the campaign for each material should be; that is, how many batches of that material should be made before changing to the next one, and how long the total wheel should be. These tools also identify products that should be made less frequently than every cycle, and products that should be made to order if lead times permit.

In addition to leveling production, product wheels bring a degree of stability and predictability to high variety operations. They provide a rational means to balance changeover costs and inventory carrying costs to optimize campaign lengths. And they provide a structured basis for calculating cycle stock and safety stock requirements, thereby allowing optimization of total inventory to that needed to support current performance.

13

Postponement in the Process Industries: Finish to Order

Postponement is an operational strategy that can be employed to maintain high levels of customer service while significantly reducing finished product inventory and WIP, and can thus be used to satisfy the lean objective of reducing inventory waste. It has greatest value where product lines have a high level of differentiation or customization—that is, where the product portfolio has the wide SKU fan out described in Chapter 2.

Industrial engineering, supply chain, and operations management texts have varying definitions of postponement. From Hopp and Spearman's *Factory Physics,* "delaying customization ... is an example of postponement, in which the product and production process are designed to allow late customization." And from Chopra and Meindl, in *Supply Chain Management,* "Postponement is the ability of a supply chain to delay product differentiation or customization until closer to the time the product is sold. The goal is to ... move product differentiation as close to the pull phase of the supply chain as possible."

Bowersox and Closs, in *Logistical Management,* describe postponement in this way: "The goal of manufacturing postponement is to maintain products in a neutral or noncommitted status as long as possible. The ideal application of postponement is to manufacture a standard or base product in sufficient quantities to realize economy of scale, while deferring finalization of features, such as color, until customer commitments are received."

From those definitions, then, postponement can be two things, one involving equipment design, and the other the way in which existing equipment can be used more advantageously:

- **Case 1:** Designing the process equipment so that the major differentiation or customization can be done at the last possible step in the process. For example, in carpet manufacture, brightly colored carpets can be tufted from colored supply yarns, or carpet can be made from neutral colored yarns, and then space dyed after tufting. A carpet mill with space dying capability can thus postpone color determination until that step in the process, rather than committing to finished product color at the raw material stage.
- **Case 2:** Given an existing process design, where the specific process step at which major differentiation occurs is predetermined, holding inventory prior to that step rather than as finished products so that the major customization can be done to satisfy specific orders.

FINISH TO ORDER

Finish to order (FTO) then is a special case of postponement, specifically Case 2. This the case of most interest to lean practitioners, because they are generally dealing with an existing production line, with equipment already in place, and are trying to improve flow and performance without major equipment replacement. So postponement strategies that rely on the location of inventory used to satisfy current demand rather than on equipment design are more useful.

It should also be noted that in process industry production, customization cannot generally be delayed to a later step in the overall process flow; the ability to customize or differentiate is often an inherent characteristic of the process equipment at a specific step. However, the customization or differentiation can frequently be delayed in time by waiting for firm orders before transforming the material through the differentiation step, in keeping with the second definition.

EXAMPLES OF FINISH TO ORDER: FTO IN ASSEMBLY

Where assembled products have a high level of final product customization, FTO is frequently practiced by maintaining inventories of all components and subassemblies just prior to final assembly. As orders are received, these components can be assembled into the specific configurations that were ordered. A frequently cited example of this is the final assembly of personal computer systems. With the variety of processors, memory size, monitors, and disc drives available, the number of possible unique configurations available from one manufacturer can run into the millions.

Figure 13.1 lists the variety of component parts that could go into the final system, which could result in millions of possible combinations. Even when one considers that not all possible combinations are likely to be ordered, that there are some preferred groupings, and that some option selections will require specific selection of other options, the number of likely possible configurations can be a few thousand. Maintaining the final inventory in subassemblies rather than complete systems reduces the total SKU count that must be inventoried to about forty.

This strategy is also frequently called assemble-to-order.

Personal Computer Options	
Option	*Number*
> Processor	3
> Operating System	3
> Hard Drive	6
> Internal Memory–RAM	4
> Monitor	3
> Printer	4
> Keyboard/Mouse	2
> CD/DVD Drive	3
> Video Card	3
> Modem	2
> Sound Card	3
> Speakers	5
Total Number of Components	41
Total Combinations Possible	1,399,680

FIGURE 13.1
Possible personal computer configurations.

FTO IN THE PROCESS INDUSTRIES

Most consumers are aware of examples of FTO with process industry products at the point of sale, although they probably don't think about it in that regard:

- At many retail gas stations, one grade of gasoline is stored in the on-site tanks, and the various octane levels are achieved by blending additives at the pump.
- Most house paint sold at hardware stores is produced by mixing a neutral base color with small amounts of color in the store, to achieve the desired shade.

Both of these are cases where a high degree of customization can be achieved with relatively low inventories, by postponing mixing until customer needs are known. This dramatically lowers the number of materials and, therefore, the total quantity of material that must be available at the point of sale.

FTO WITHIN PROCESS PLANTS

Although those are the examples of process industry postponement that consumers are likely to be most familiar with, the majority of postponement occurs within the plant walls.

- Crop protection products, such as fertilizers, pesticides, and fungicides, are produced in a relatively small number of formulations, but have tremendous variety in packaging configuration. Some of the differentiating factors are package size and weight; package construction, with moisture barriers built in for products going to humid climates, for example; and labeling requirements for products being sold into global markets. Holding material in bulk and then packaging to order can dramatically reduce inventory requirements.
- In the manufacture of plastic pellets, to be sold to manufacturers who will then mold them into parts of various shapes and sizes, for

end uses ranging from soda bottle caps to seat belt retractor mechanisms, the pellets are typically extruded and stored in silos. In some cases, they are taken from the silos, compounded with pellets having different properties, and re-extruded. Thus, pellets with a wide range of properties can be made from a few basic pellet types. It is advantageous from an inventory standpoint to maintain stocks of base pellets in the silos, and then compounding to satisfy specific orders, rather than carrying finished product inventory of compounded pellets in all possible combinations.

- As will be seen below, our sheet goods process has 2,000 slit and chopped finished products, but only 200 in the pre-slit form. So if we can hold inventory prior to the slitters, and then slit, chop, and package to order, the number of types that must be inventoried is reduced by a factor of ten.

THE BENEFITS OF FTO

The benefits to be gained from moving to a FTO strategy are numerous. Forecasts tend to be much more accurate at the aggregate level than at the finished product level, so to the extent that forecasts are used to determine inventory requirements, inventory kept at the FTO point will be based on much more reliable information.

Although the inventory stored at the FTO point may have to increase, finished product inventory can be eliminated almost entirely. Safety stock, the component of inventory needed to protect delivery performance against variability, will typically be reduced by the square root of the reduction in SKU count, so if the number of varieties at the FTO point is one tenth of the number of final product SKUs, the required safety stock will be reduced by the square root of 10, a factor of about 3.1. It is generally the case that any increase in WIP at the FTO point is far more than offset by the reduction in finished product inventory, so total inventory in the system is significantly reduced.

The highly differentiating steps after the FTO point will be performed based on actual customer orders, so asset capacity will not be consumed making products that might sit in final inventory for long periods of time. Because assets are not occupied with making currently unneeded products, they are more available to make products that are being ordered, so customer service level often improves, in spite of the absence of finished product inventory. The following list summarizes these benefits:

- Reduced dependence on forecasts
- Highly differentiating steps performed based on true demand
- Reduced finished product inventory
- Reduced total inventory
- Final process steps more available to produce to immediate needs
- Improved customer service levels

EXAMPLE OF FTO IN A PROCESS PLANT

To walk through a specific example of what FTO would look like in a process operation, the sheet goods process described in Chapter 2 and several subsequent chapters will be revisited. Figure 13.2 shows a portion of the future state value stream map (VSM), after the cellular manufacturing concepts of Chapter 11, and product wheels as described in Chapter 12, have been implemented. The concept to be applied here is to hold enough inventory in the WIP storage between the bonders and slitters, and then do all slitting, chopping, and packaging to real customer orders (Figure 13.3). When an order is received, it will be put on the schedule for the appropriate slitter (as defined by the cellular product lineup). Once slit, it will flow to the pre-chop WIP; because the choppers are not dedicated to specific cells, the three choppers each serve all three slitters, so some inventory is needed to buffer rolls waiting to be chopped. The delay waiting for the appropriate chopper will typically be a few hours, always less than a day. Once chopped, the rolls flow through packaging and labeling, to be loaded onto trucks to be shipped to customers.

So the primary inventory from the bonders forward becomes the inventory in front of the slitters. No more inventory is needed here than before because the product wheel design provided enough inventory that all bonded styles are available in the quantities needed to immediately satisfy customer orders. But almost all finished product inventory has been eliminated; the only finished product inventory is that being staged temporarily for truck loading. As can be seen from Figure 13.3, the future state VSM after

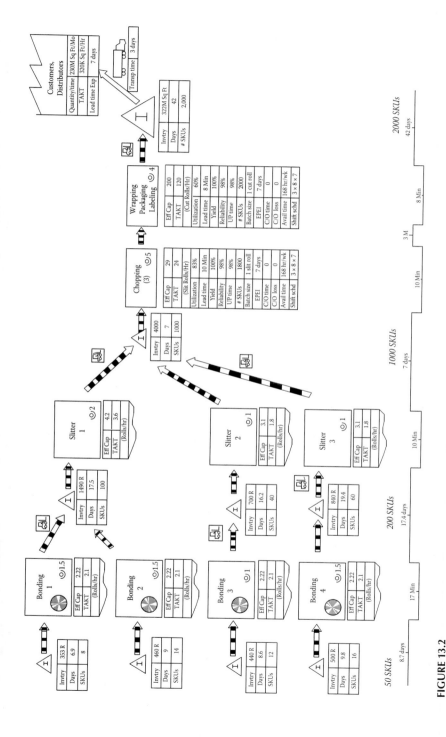

FIGURE 13.2

Sheet goods process (Future State 2, with cells and product wheels).

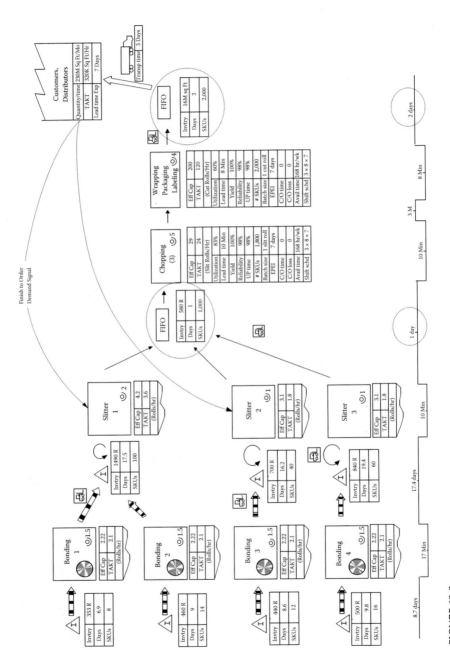

FIGURE 13.3

Sheet goods process (Future State 3, slit to order).

implementing slit-to-order, we now need a total of 20.4 days (17.4 + 1 + 2) of inventory downstream of bonding compared to 66.4 (17.4 + 7 + 42) days before.

Thus, the total inventory in the back end of the plant (bonders through shipping) has been reduced by 69 percent!

There are three requirements that must be met at the FTO point:

1. The total lead time from the FTO point through shipping must be less than the customer lead time commitment, as is explained in Chapter 12. That condition is satisfied here, because our commitment to customers is to deliver within seven days of order receipt. The manufacturing lead time from the FTO point out can be seen from the timeline at the bottom of the VSM in Figure 13.3, and is about three days. One day of this is the maximum time that rolls could spend waiting for the appropriate chopper, and the time to stage rolls and load the truck is two days. The slitting, chopping, and packaging times are all at ten minutes or less. Transportation to the customers takes a maximum of three days, so the seven-day customer delivery expectation can be met with one day to spare.

2. The process steps downstream of the FTO point must have sufficient capacity to meet the short-term demand placed on them. The slitters, choppers, and packaging systems all meet this requirement.

3. The product must be in a physical form that permits storage at the FTO point. Here, everything is wound on rolls, which can readily be stored. In a plastic pellet operation, the pellets are generally easy to store, but not so earlier in the process when it is in the molten polymer form. In a food processing operation, formed but unbaked dough probably cannot be stored for any length of time, so any FTO point would have to be upstream or downstream of that point in the process.

A FURTHER EXAMPLE: BOND TO ORDER

It would be interesting to investigate taking the slit-to-order strategy to the next level, that is, to bond-to-order. Figure 13.4 depicts the future state VSM for this operational strategy. Most of the inventory between bonding

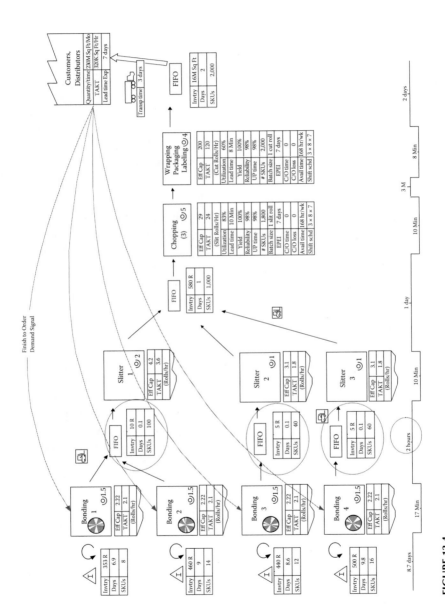

FIGURE 13.4

Sheet goods process (Future State 4, bond to order).

and slitting can be eliminated, and the primary inventory required to satisfy customer orders is now ahead of the bonders. The inventory immediately following the bonders has been reduced from 3,030 rolls to only 20 rolls, with the days of inventory down from 17.4 days to two hours!

But, there is a problem! As noted earlier, customers expect product to be delivered within seven days of receipt of order. With this strategy, the manufacturing lead time from the FTO point out is ten days; adding shipping time brings the total lead time to thirteen days. Each bonder is running a one-week product wheel, so the lead time through the bonder for a specific SKU can be as high as a week; if the bonded style required has just completed its spoke, almost a week will elapse before that spoke comes around again. So the total manufacturing lead time is seven days for the bonder product wheel, plus one day for the pre-chop inventory and two days to stage, load, and ship. That means that this future state cannot be implemented with today's performance.

However, the potential inventory reduction is very large. A reduction in WIP of 3,010 rolls represents 150 million square feet, so the inventory savings and the carrying cost savings are several million dollars each. Even though this future state cannot be immediately reached, it should be a goal, with activities chartered to improve the specific performance levels required to shorten total lead time to seven days or less.

- Better scheduling and flow between the slitters and choppers can dramatically reduce the one-day delay there.
- Improvements to the finished product picking, staging, and truck loading processes should significantly reduce that two-day portion of the lead time.
- The product wheels on the bonders should be reexamined to understand the factors that led to the selection of one week as the optimum. Perhaps a concentrated SMED activity can reduce the forty-five-minute changeover time, which likely was a key influence on the original product wheel design.

Each of these improvement activities can be done as a kaizen event, as described in Chapter 9.

The key point here is that desired improvements in operational strategy such as bond-to-order should not go undone because current performance won't allow it. The desired future state should be used to drive process and flow improvement. And as the goal future state is reached, the next level of future state should be defined. A healthy continuous improvement culture should always have a currently unsatisfied future state VSM to be working toward.

SUMMARY

FTO is a strategy that can be beneficial in many situations, specifically in processes with a high degree of product differentiation, so consider it whenever you encounter that situation. Where manufacturing lead times allow it, it can reduce inventory dramatically, as can be seen from the slit-to-order example. Where current manufacturing lead times don't allow FTO, but the potential benefits are great, as in the bond-to-order case, programs to reduce lead times should be aggressively pursued.

The ultimate goal should be to move the FTO point back far enough toward the beginning of the process that it comes before any highly differentiating steps, where product variety is still low. Where that can be achieved, inventory will be needed in only a small number of varieties, so total system inventory can be reduced to the bare minimum.

14

Pull Replenishment Systems

Pull replenishment systems are just as applicable to process plants as they are to assembly plants, and can bring equal benefit. But before discussing how to apply pull to process lines, it is important to clarify just what pull means. The lean literature offers some confusing, and often conflicting, definitions of pull. Some imply, for example, that a make-to-order strategy is inherently pull, while others state that it is inherently push.

WHAT IS PULL?

Most references seem to agree that pull includes replenishment modes where production is allowed whenever material has been consumed—or pulled—from a downstream inventory, and that the need to replace this pulled material is conveyed by some visual, real-time means. However, there is not universal agreement on situations where production is allowed, not to replace consumed material, but to respond to a signal that a customer or downstream operation needs material immediately.

One reason for the lack of consistent definitions is that Taiichi Ohno, considered by most to be the father of pull, did not explicitly define pull. He describes it in a way that supports the replenishment-only school of thought:

> Manufacturers and workplaces can no longer base production on desktop planning alone and then distribute, or *push,* them onto the market. It has become a matter of course for customers, or users, each with a different value system to stand in the frontline of the marketplace, and so to speak, *pull* the goods they need, in the amount and in the time they need them.

So this can be interpreted as saying that as customers pull goods from a finished product inventory, production is authorized to replace them. The basic kanban strategy reinforces this mind-set. Thus, a number of authors describe pull in a way that includes only replenishment of goods consumed. Hopp and Spearman, in *Factory Physics*, explicitly state that pull systems "authorize production as inventory is consumed." And "Another way to think about the distinction between push and pull systems is that push systems are inherently *make-to-order*, while pull systems are *make-to-stock*."

Other authors take a somewhat broader view, that pull includes both replenishment of consumed stock and production in response to a signal from a downstream operation that more material is needed now. Art Smalley, in *Creating Level Pull*, treats both situations as pull, and uses the terms "replenishment pull" and "sequential pull" to distinguish them. Masaaki Imai, in describing pull in his book *Gemba Kaizen*, says, "the entire plant springs into action with the receipt of an order from a customer."

Before deciding on which definition is more appropriate, it would be helpful to consider the reasons that Ohno and Toyota developed pull. The primary reason appears to be to avoid the waste of overproduction, that parts "reach the assembly line at the time they are needed and only in the amount needed." Other objectives Ohno had in mind were to synchronize flow, and to reduce the material management tasks associated with production planning and scheduling.

It seems logical, therefore, to broaden the definition of pull to include all models that avoid overproduction by explicitly limiting inventory to the minimum needed to ensure smooth flow of products to customers, synchronizes that flow to customer needs, and communicates those needs in a highly visible manner. A suggested definition of pull: *A system that produces to replenish material that has been consumed, or material for which there are firm orders needing to be filled immediately, and in which flow is managed and synchronized by current conditions in the operation.*

By that definition, make-to-order can be pull. It can also be push, if flow and inventories are not managed in a way that limits WIP buildup. Likewise, make-to-stock can be push or pull, depending on how inventories are managed. One slightly imperfect, but useful test is to ask if current production is scheduled based on a forecast; if so it is almost always push. If current production is based on current conditions on the plant floor, it is most likely pull.

PULL IN ASSEMBLY

The vast majority of documented cases of pull production are in assembly processes. Pull has brought great value to these processes for the very reasons that Ohno and his contemporaries at Toyota developed it: It reduces or eliminates overproduction, synchronizes flow to current demand, and simplifies all the managerial tasks associated with production scheduling.

Many of these applications employ kanban logic, which forms the communication chain flowing in reverse through the process, starting with customers and flowing back all the way to raw materials. Kanban literally means a signal, or a visible sign. In typical systems, the signal can take the form of cards, totes, or spaces marked on the floor which, when empty, signal need to replenish.

The basic kanban concept is shown in Figure 14.1 and Figure 14.2. In this example, it has been determined that four containers of finished

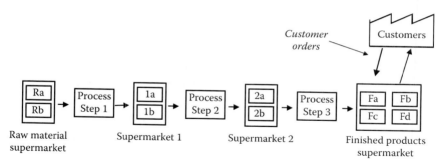

FIGURE 14.1
Kanban concept.

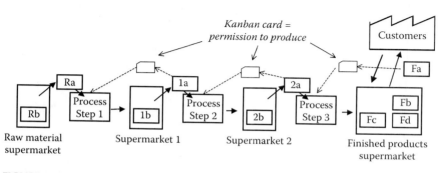

FIGURE 14.2
Kanban operation.

product is the appropriate inventory to satisfy customer demand for one specific SKU in a timely fashion, so the finished product supermarket has four containers with that product. (Note that the term "supermarket" is often used to describe inventories managed by kanban principles. Indeed, Ohno first conceived pull and kanban processes after studying American supermarkets in the late 1940s. He was impressed that in a supermarket, customers can get what they need, in the quantity they need, when they need it, and that the shelves were restocked with only what customers had pulled from the shelf.)

It has also been determined that two containers of WIP are required between each process step to ensure smooth flow through the plant. When a customer orders a container (see Figure 14.2), one container (Fa) is removed from inventory and shipped to that customer. Before it is shipped, a card on the container is removed and sent to the final processing step (Step 3) to signal permission for that step to produce another container of that material to replace the one the customer has pulled. In order to produce a replacement, Step 3 needs input material, so it pulls a container of WIP (2a) from Supermarket 2. A card on that container is removed and sent to Step 2 to signal permission to replenish. In that manner, as material is moving from left to right through the process, the kanban signals, in the form of cards, are moving from right to left, or upstream in the manufacturing process.

Instead of cards, this system could use empty containers as the kanban signals. When the material in container Fa is shipped to the customer, the empty container could be sent back to process Step 3 to signal a need to replenish. After Step 3 has pulled WIP Container 2a and consumed the material, that container would be sent back to Step 2 to permit production to refill it.

Alternatively, the system could use empty container spaces in the supermarket to signal that a container has been pulled and needs to be replaced.

These are all examples of kanban, and demonstrate both the signaling feature and the fact that production only occurs to fill a need to replace pulled material. Thus, the objectives of limiting overproduction and flow synchronization are satisfied. Various authors describe different types of kanban to provide different functions. Among the types commonly mentioned are production kanban and withdrawal kanban; the designations temporary kanban, strategic kanban, and trigger kanban are sometimes mentioned. Ohno describes withdrawal kanban, transport kanban, production ordering kanban, and signal kanban.

Those distinctions are not important to this discussion. What is important to understand about kanbans and supermarkets:

- Some inventory is needed by most processes to maintain flow that matches takt.
- In a pull system, inventories must be managed in a way that prevents overproduction.
- Kanban signals are one effective way to mange flow and inventories.
- Kanbans are not the only way to achieve pull. An alternative, called ConWIP, is described later.
- In any case, there must be a communication or signaling process that is visible to all involved in real time replenishment decisions; that is, anyone who gives or receives permission to produce.
- Production is triggered either by consumption of material in inventory or by a demand signal from the downstream operations, not by a schedule created from a forecast.

DIFFICULTIES IN PROCESS PLANTS

Process plants pose two unique challenges that must be met for pull to be feasible.

The first challenge is that pull inherently requires that equipment be stopped and started within short periods of time. When a given step in a process has produced enough to meet the immediate need, whether signaled by kanban or by some other signal conveying an immediate need, the step must stop or material will be produced that is not currently needed. This is one of the most fundamental characteristics of any pull system, that production stops as soon as the need has been satisfied. A difficulty arises when production equipment cannot be stopped, which is not uncommon in the process industries. Many plastic pellet, synthetic fiber, and paint manufacturing operations begin with a polymerization process, which takes various chemicals, mixes them, and then reacts them at high temperatures and pressures, so that the molecular chain length will grow and properties such as tensile strength and viscosity will increase. Continuous polymerization processes are very difficult and costly to stop. In many cases, if flow stops even for minutes, the polymer will solidify in the vessels and piping, requiring that everything be disassembled and burned out before the process can be restarted. It is not uncommon for these processes to run for twenty-four to thirty months before stopping for a process overhaul.

Other chemical processes are so costly and time-consuming to restart that shutdowns are planned months in advance, regardless of whether the shutdown is to prevent overproduction or to perform necessary maintenance.

These are extreme examples, but there are many process industry systems, although not that severe, that don't lend themselves to starting and stopping quickly in response to the presence or absence of pull signals.

A second challenge when applying pull in process plants is the large number of product types in the latter stages of production. If inventories are managed by kanban in a process with 200 semi-finished types, there must be kanbans for all 200 types, which can be a large amount of inventory, thus defeating the primary purpose of pull.

Fortunately, there are techniques for addressing both of these challenges.

PUSH–PULL INTERFACE

When designing a pull system for a process with a step that can't readily be stopped, creating a push–pull interface is often a reasonable compromise. In many cases, this offers much of the benefit of pull, without incurring the extreme cost of stopping the challenging pieces of equipment. The concept is to find a point in the process downstream of all difficult-to-stop steps, and where inventory can conveniently be maintained, and designate it as the push–pull interface. Material is then pushed through the earlier process steps to accumulate in the inventory at the push–pull interface. Material is then pulled from this inventory only as downstream needs are signaled.

This works well in many process operations because the steps with difficulty in stopping are usually early in the process, where little product type differentiation has taken place. Forecasts are generally more accurate at this aggregate level, so the mix can be adjusted so that the materials being pushed are likely to be the specific types currently needed. In other words, if you must overproduce, it is better to overproduce materials that are likely to be needed sooner rather than later. At the downstream steps, where most of the differentiation typically occurs, material is being pulled, so the product types being made are those needed immediately.

As an example, consider the plastic pellet manufacturing operation shown in Figure 14.3. Raw materials are mixed and then polymerized in the continuous polymerization step. The material is then extruded and cut as pellets, which are stored in large silos. Two or more pellet types are conveyed from the silos into blenders where the different pellet types are mixed and then fed to compounding extruders. The blended and extruded pellets are fed to small silos to be stored prior to either boxing or bagging. Customer orders are filled from finished product inventory kept in an off-plant warehouse. In the current state, everything is push. Each step is scheduled independently of the others, all from forecast information. Consequently, inventories are high.

The continuous polymerization systems and the first set of extruders cannot be stopped, but the downstream steps, the blenders, compounding extruders, and packaging lines can be easily started and stopped with little advance notice. These are batch operations, so the current batch must be completed, but the next batch can easily be postponed. Thus, it is possible to designate the large silos as a push–pull interface (Figure 14.4). Material is pushed through polymerization and primary extrusion into the large silos. The small silos and finished product warehouse are now managed as pull supermarkets. The flake blenders and compounding extruders are no longer scheduled by forecast; they produce in response to a signal that material has been pulled from the small silo supermarket. Likewise, packaging operates only in response to pull signals from the warehouse supermarket.

Thus push, and the opportunity for overproduction, exists only in the first steps, where there are few types (eight), so product mix is far more predictable. Pull in the downstream steps eliminates overproduction, so the inventories are significantly lower, and flow is synchronized to actual customer orders. In this example, total inventory in the system has been cut in half, from a total of 116 days down to 50 days.

As mapped in Figure 14.4, this is not the FTO concept introduced in the previous chapter. In this case, we are not finishing to order; it is a full MTS strategy. It is easy to confuse push–pull interfaces with FTO, because they can occur at the same point in the process, and are often done together. But they are two different things. With a push–pull interface, upstream steps are scheduled from a forecast and downstream steps from current needs. With an FTO point, upstream steps may be pushed or pulled; downstream

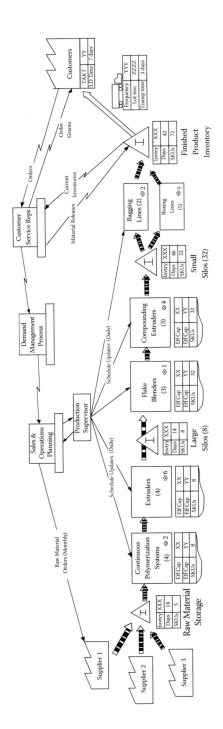

FIGURE 14.3

Pellet manufacturing current state VSM.

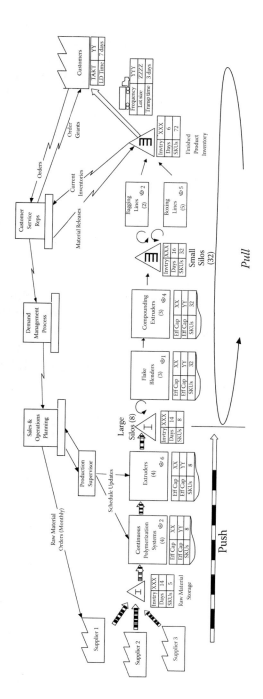

FIGURE 14.4

Push–pull interface concept.

steps are generally pulled, but may be pushed if downstream inventories are not appropriately managed. Another way to think about the difference is that the push–pull interface is selected based on which process steps are readily capable of starting and stopping and which steps are not, while the FTO point is governed by the customer lead time expectation and how far back into the process the manufacturing lead time allows material to be finished without exceeding it.

With the pellet process from Figure 14.3 and Figure 14.4, we could FTO from the push–pull interface, if the lead time through the blenders, compounding extruders, packaging, and shipping is less than the customer lead time promise. If not, FTO may be possible from the small silos. Figure 14.5 depicts this situation. Here, material is pulled from the supermarket in the small silos in response to customer orders, and sent to packaging. The finished product supermarket has been replaced by a FIFO (first in – first out buffer); the bagged and boxed material flows through that area, and stops only long enough to get a truck staged and loaded. So rather than pulling material from the small silo supermarket to replenish the finished product supermarket as depicted in Figure 14.4, material is pulled from the small silo supermarket in response to customer orders. And because finished product is stored in FIFO fashion only long enough to get loaded on a truck, inventory at that location is reduced from six days to less than one day.

Thus, in this case, there are separate push–pull interface and FTO points, at different points in the process. In many cases they do coincide.

The push–pull interface is not a perfect solution to the nonstop nature of some process equipment, but is a reasonable compromise that supports many of the benefits of pull in those situations where equipment can't stop.

CONWIP

ConWIP (constant work in process) is a strategy for managing production that limits in-process inventory and can be used as an alternative to kanban. ConWIP can control an entire process, or several steps in the process. The fundamental idea is that nothing is allowed to enter the process segment being controlled by ConWIP, the ConWIP loop, until something leaves. This one-for-one relationship between lots leaving and lots entering

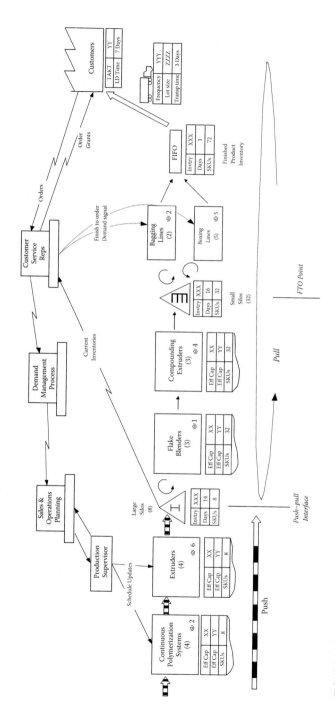

FIGURE 14.5

Push–pull interface with FTO point.

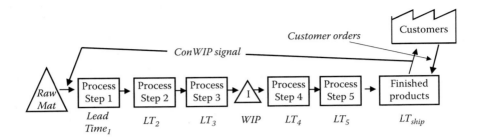

FIGURE 14.6
ConWIP in a make-to-stock environment.

maintains work-in-process within the loop to a fixed amount. Thus, over-production is prevented, and ConWIP can be used as the control mechanism in a pull system.

Hopp and Spearman, in *Factory Physics,* demonstrate that a ConWIP system is less sensitive to errors in WIP level settings than a push system is to errors in setting throughput. They also point out that ConWIP systems are inherently easier to control; they require fewer cards in circulation versus a kanban system. And regardless of the specific signaling technique being used, fewer control signals are needed.

Figure 14.6 illustrates the ConWIP concept being used to control the entire process in a MTS environment. As customer orders are received, they are filled from finished product inventory. The material leaving the warehouse generates a signal that Step 1 is permitted to begin to make another lot of that specific material. As more orders are received, and material shipped to satisfy them, lots of those materials are allowed to begin their way through the process.

One of the most significant features of ConWIP is that the only material in any WIP storage within the process is in the materials currently being processed. In a process industry plant where hundreds or thousands of final SKUs can be produced, the WIP within the operation at any time is only in a fraction of those. In contrast, in a kanban controlled system, WIP must include kanban quantities for all product types that can exist at that point. Thus, ConWIP can overcome the process industry challenge mentioned earlier, by requiring only reasonable amounts of WIP even in highly differentiating processes. For that reason, ConWIP has sometimes been called *part-generic kanban* in contrast with the more traditional idea of part-specific kanban.

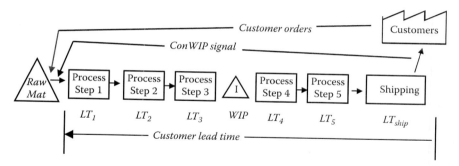

FIGURE 14.7
ConWIP in an MTO environment.

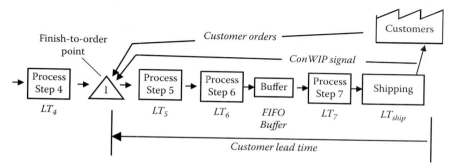

FIGURE 14.8
ConWIP in an FTO environment.

ConWIP can also be used to control an MTO strategy, where there is no finished product inventory, as shown in Figure 14.7. As customer orders are received, they are queued up in front of Step 1, and as each previous order is shipped, process Step 1 is given permission to pull the raw materials required for the next order in the queue and begin to process them. As before, the one out/one in feature limits WIP to a fixed level, and WIP is only needed in the types of materials currently being processed.

Note that in an MTS environment, the next material to be processed is the same material as just left. In contrast, in an MTO environment, the next material to begin production is the material for the next order that was received.

ConWIP being used to control a segment of a process using an FTO strategy is shown in Figure 14.8. Here the inventory between Step 4 and Step 5 forms the FTO point. As in the prior example, as customer orders come in, they are queued up in front of Step 5. As completed orders are shipped, that generates a permissive signal for Step 5 to begin finishing the next order in the queue. In any of these examples, any inventory within the ConWIP loop is managed as a FIFO buffer, not as a supermarket.

DEVELOPMENT OF PULL ON THE SHEET GOODS PROCESS

Now we explore how these concepts apply to the sheet good forming process described and improved upon in prior chapters. Based on the work described in Chapter 11, there are four virtual work cells, each containing a forming machine and a bonder. In Chapter 12, product wheels were designed for each forming machine and for each bonder. The forming product wheels are based on changeovers prioritized in three ways: by sheet width; by basis weight; and by polymer type. In Chapter 13, an FTO strategy was implemented at the slitters. (Remember that bonding-to-order was studied, but the long manufacturing lead time prevented it from being implemented until the bonder product wheel time could be reduced.)

That FTO strategy (slit to order) was depicted in Figure 13.3. The post-slitting and post packaging inventories are managed as FIFOs, buffers with short residence time, with only enough inventory to enable chopping to synchronize with slitting, and to allow for product staging for truck loading.

The flow strategy at the back end may look somewhat like ConWIP in that customer orders are loaded onto the slitter schedule, and that all downstream WIP is managed as FIFO rather than supermarket. But it is not true ConWIP because there is no one-to-one correspondence between shipments and new orders starting on the slitters; each slitter takes the next order in its queue when it has time available, independently of the timing of orders being shipped. Thus, WIP is not explicitly capped; the slitters may run slightly faster than truck loading at some times, so the FIFOs may increase temporarily. However, any such increase will be temporary. Thus, flow through the slitters, choppers, and packaging is pull by the definition given earlier, as production is in response to a specific customer demand signal.

This strategy is like ConWIP in that there are few SKU types in the FIFO buffers at any time. Only the specific rolls waiting to be chopped, or waiting to be loaded onto a truck, are in the buffers. In contrast, if those inventories were managed as supermarkets, there would need to be 1,000 and 2,000 different materials, respectively, in the two supermarkets. Thus, the inventory in the back end of the process is dramatically reduced.

As the process is depicted in Figure 13.3, the inventory before slitting becomes a push–pull interface: As the process is currently being managed, material is still being pushed through forming and bonding. With this process, however, there is no need for a push–pull interface. Because both the forming machines and the bonders can be started and stopped without significant penalty, the entire process can be put on pull flow.

A full MTO strategy cannot be employed. The lead times through forming and bonding cause the total manufacturing lead time to be much longer than customer lead time expectations, as was discovered earlier, so the FTO point prior to slitting will be kept for now. Future improvements to the bonding and forming product wheels may allow the FTO point to be moved upstream in the process, and may eventually permit a complete MTO strategy.

Figure 14.9 shows the process on complete pull. The inventories feeding the slitters, at the FTO point, are now supermarkets. Raw materials (RM) are now on pull; orders placed with suppliers are now based on the quantities that have been pulled from the RM supermarket by the forming machines. RM inventory has accordingly dropped from 16 days (6.3 million pounds) to 3.1 days (1.2 million pounds). See Appendix A for the calculation.

Inventories between forming and bonding are also supermarkets. Note that inventory has not been reduced; product wheel times must be reduced to do that, and that may require reduction in changeover time and cost.

As a further refinement to the pull strategy, we can treat each forming-bonding cell as a ConWIP loop, as shown in Figure 14.10. The factor that makes this feasible in this specific process is that with the product allocation resulting from cell design, the optimum sequence on each bonder is similar to the optimum sequence on the forming machine in its cell. So in each virtual cell, the forming and bonding product wheels can be synchronized so that bonding is always processing rolls that were formed a short time (a day or less) ago. The specific operation within Cell 3, for example, is as follows. When Slitter 2 pulls a roll from the pre-slit supermarket to satisfy a customer order, that causes a ConWIP signal to be

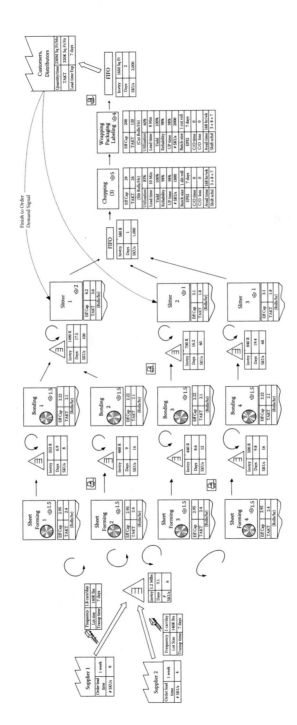

FIGURE 14.9

Pull in the sheet forming process.

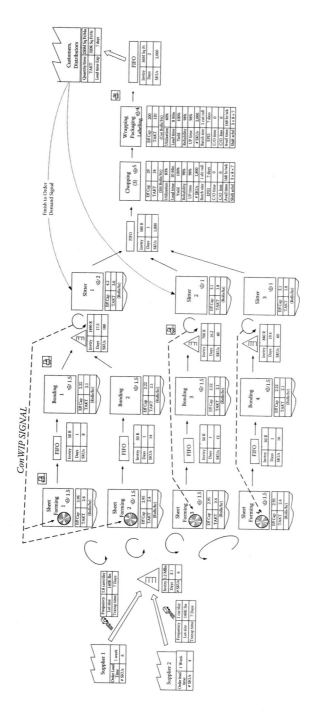

FIGURE 14.10

Sheet forming—pull using ConWIP.

sent back to Forming Machine 3 to produce another roll of that material. Forming Machine 3 loads that request onto its product wheel schedule, and when the appropriate spoke comes up in the wheel sequence the roll gets formed. The roll flows into the FIFO, and is bonded within a few hours. The bonded roll flows into the pre-slit supermarket to replace the specific roll that had been pulled earlier.

In reality, bonding is no longer running an independent product wheel; it is following the wheel on the forming machine.

This is true ConWIP in that forming is permitted to produce rolls only in one-for-one correspondence to rolls that have been pulled from the pre-slit supermarket. It does not produce them *in the same sequence* as they have been pulled; the sequence is set by forming product wheel criteria. There may be some delay in the rolls getting formed because of the wheel cycle, so WIP is not truly constant, but it is explicitly capped. It may be less than the cap at times, but never greater. The benefit of ConWIP in this situation is that the FIFOs within the loops contain only the types of rolls currently being processed, whereas the supermarkets at those locations had required all formed SKUs (a total of 50) to be stored in sufficient quantity to meet any reasonable demand. Thus, the total inventory between forming and bonding has been reduced from 1,753 rolls down to 200, an 88 percent reduction!

An additional benefit of this ConWIP loop is that rolls are always bonded "fresh" (e.g., within a day of being formed). Although not a requirement, the bonding process tends to perform better, and gives slightly higher yields, with fresher rolls.

VISUAL SIGNALS

Visual signals are an extremely important part of any pull system. Kanban, the heart of Toyota's pull system concept, literally means visible sign. Although card systems are often the signal of choice in assembly operations, and have seen successful application in process operations, the takt boards described in Chapter 8 have proven to be a very effective way to satisfy this need. Takt boards have seen widespread use in process plants, in a variety of different types of processes, ranging from the manufacture of plastic pellets to the production of X-ray films.

To get a more detailed understanding of how takt boards provide the visual communication, consider a process line making plastic materials that customers mold into gears. The line runs a seven-day product wheel, which begins every Wednesday morning, at 8 a.m. On Tuesday morning, the production planner checks the current inventories of all materials made on that line. From that, she determines what must be made on the next wheel cycle to return inventories to the cycle stock plus safety-stock target. What she will schedule to be produced will typically not be exactly the cycle stock. If the inventory data show that some of the safety stock has been consumed on the current cycle, she will schedule replacement of that safety stock in addition to the cycle stock. If, on the other hand, not all of the cycle stock has been consumed, that there is more than the safety stock remaining, she will schedule production of less than the cycle stock. (The fact that there is an offset in time from the day on which the inventory is checked (Tuesday) and production of a specific material will be started (say, Friday) constitutes what some operations management texts call a "review period." This can be easily accommodated, and is covered in Chapter 15.)

Once the quantities to be produced have been calculated, the production plan is set. The sequence in which the various products are to be made had been determined as part of the product wheel design. The production plan will then be communicated to the person responsible for running the line. This may be an area supervisor, or may be a flow manager as was described in Chapter 11. In this case, communication of the plan is done electronically because the planner is located in an office several hundred miles from the plant.

The flow manager checks the plan to make sure that it accommodates any special events, such as production of a test product or a planned outage. He then meets with a member of the production team, who puts the schedule on the takt board, which in this case is handwritten but could be electronic. Thus, the board provides a visual signal to everyone in the area as to what is to be produced to replenish what has been consumed from inventory. It has the added benefit of providing a place to record production to the plan, and to log production difficulties, as described in Chapter 8.

In the case of the sheet goods pull system, all signals are conveyed electronically. As a roll is pulled from the pre-slit supermarket, for example, an electronic signal is sent to the system maintaining forming product wheel requirements, to tell it that another roll of that material should be formed. The current forming product wheel schedule is displayed on a large computer-driven screen located in the forming area.

WHEN TO START PULLING: THE SEQUENCE OF IMPLEMENTATION

There are at least two schools of thought on when it is most appropriate to begin to implement pull on part or all of a process. One school believes that getting to pull is an essential component of any lean journey, and that pull is the platform to drive continuous improvement. Pull should, therefore, be implemented soon, and it will drive the desired process flow improvements.

The other school teaches that pull is better done after other flow improvements have been accomplished, after the process is flowing smoothly, with high stability and discipline, and all process variability has been reduced as much as possible. This school's belief is that pull will be easier to design and install if the process has been simplified and stabilized, and will be easier to sustain if there are fewer interruptions and special cause events.

The latter view is somewhat reinforced by Umble and Srikanth in *Synchronous Manufacturing:* "One weakness of the JIT approach is the inability to identify systematically the critical, capacity constraint resources in the operation in advance. The Japanese approach to attacking waste and supporting the process of continuous improvement within the plant is essentially unfocused."

Although I have seen successful implementations done following each concept, I have a strong bias toward the second view. I strongly recommend a bottom-up approach, as shown in Figure 14.11. Any lean effort should start with creation of a VSM, and then focus on some of the basic tools as a way to engage the entire workforce, including 5S, standard work, SMED, and TPM. Once the foundation has been laid, it is time to work on flow improvement and bottleneck management. At this time, all causes of variability in the process should have been minimized. That sets the stage for virtual cell design (if the equipment configuration is appropriate for cells), and then product wheels. It is much easier to implement product wheels in processes with parallel assets if the flow has first been aligned within cellular boundaries. With all of that accomplished, everything has been optimized in a way that gives pull the highest likelihood of success.

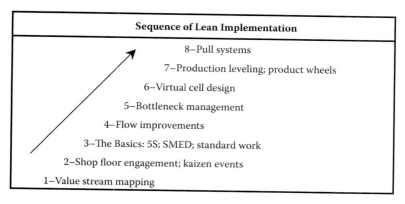

FIGURE 14.11
Implement from the bottom up.

Pull systems are far more practical if production has been leveled, using *heijunka* concepts such as product wheels.

The opposite approach, starting with pull and using that as the platform to identify and prioritize flow inhibiters, often encounters so many issues that results are not as expected, so people get discouraged, and the will to continue fades.

Of course, all these should be done with business strategies and policies clearly defined, and be based on targets set by the business (e.g., customer lead times and service levels), as described in Chapter 16.

CREATING PULL

To summarize the preliminary tasks that should have been done before pull design begins:

1. Ensure that you have a thorough, accurate VSM
2. Optimize flow; reduce process variation
3. Ensure that all bottlenecks have been identified, optimized and that constraint management principles are being followed
4. Implement cellular manufacturing where appropriate; load products onto parallel assets in a way that
   - minimizes products being made on more than one asset
   - aligns families with assets to minimize transitions

5. Decide whether product wheels are appropriate; if so, ensure that wheel design is up to date, with
   - Optimum sequence
   - Optimum cycle
   - Optimum frequency and spoke length for each product

The specific tasks required to implement pull, once all the preliminary tasks have been done, are:

1. Analyze the start-stop practicality of each asset and note the turn-down ratio of each. Sometimes equipment that can't be stopped can be reduced in rate enough that it can support pull flow rules. Decide whether you need a push–pull interface point. If so, decide where to place it.
2. Decide which products will be make-to-order and make-to stock. Decide on any FTO points.
3. Finalize the end-to-end pull concept. Consider where ConWIP might be advantageous.
4. Decide where inventories will be needed, and designate as supermarkets, FIFO buffers, or push inventories.
5. Calculate required inventories, the maximum and average values. (See Chapter 15 for specific methods to calculate supermarket inventory levels.)
6. Design the replenishment signals.
7. Design and construct the visual management tools, using the concepts described in Chapter 8.
8. Consider whether a computer simulation of the pull operation would be informative. Discrete event simulations can be used to:
   - Optimize supermarket levels
   - Predict customer service levels
   - Build confidence in the design
9. The supermarket inventory levels will generally be quite different from the current state, sometimes higher and sometimes lower, when analyzed on an SKU by SKU basis. Develop an inventory ramp-down/ramp-up plan.
10. Implement the changes required for the ERP system (e.g., SAP, MIMI, Manugistics).

11. Turn it on.

12. Monitor supermarket levels and fine-tune them as actual experience is gained.

VALUE STREAM FOCUS

It must be emphasized that what is being pulled in a pull system is the value stream, not the equipment. Lean implementers sometimes say that they want to put an extruder, or an autoclave, or a carpet tufting machine on pull. What they mean to be saying (or should mean) is that they want to put the value stream flowing through that piece of equipment on pull. This is an important distinction, and far more than just semantics or terminology.

Figure 14.12 illustrates a situation where some of the equipment, but not all, is intended to be put into a pull replenishment mode. This is being done as a pilot, to prove the benefits of pull, to get everyone trained and accustomed to a pull mode of operation, and to fine-tune the concept before taking pull across the entire process.

In some cases this is a sound strategy, and will lead to a smooth overall implementation. In other cases, it is a mistake and will result in confusion and failure. The difference lies in product assignment, in whether the value streams flowing through Cell 1 are dedicated to Cell 1 or can also flow through Cell 2. If any value streams can flow through both cells, they will be pulled at some times and pushed at others. Thus, the permission-to-produce pull signals will be difficult to generate and will be confusing. This sharing of value streams across parallel assets is common in the process industries, so the potential for this situation occurring is real. If lean practitioners are thinking in terms of equipment, not in terms of value streams, the potential is greater.

In situations where piloting pull on one cell or on one major piece of equipment is the best way to begin pull, the problem can be resolved if all value streams flowing through the pilot can be dedicated to that piece of equipment or cell. If not, both cells should be included in the pilot so that specific value streams will be on pure pull.

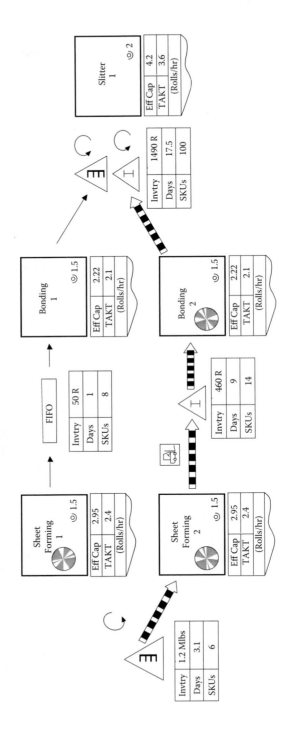

FIGURE 14.12
Pull value streams (not assets).

TRADITIONAL PULL STRATEGIES AND SIGNALS

The process plant cases described thus far may make it seem like the more traditional kanban systems are not useful in process operations. That is certainly not the case; there are many examples of these approaches being used to great success in process plants.

A typical example is the paint manufacturing operation that was depicted in Chapter 11. A polymerization reaction process makes the base resins, which are placed in large (250 gallon), stainless-steel portable totes. These totes are stored until a specific type is needed by the high-speed dispersion process. As each tote is filled at the output of the polymerization process, a card is placed in a slot on the tote. When the tote is pulled from storage to be taken to dispersion, the card is removed and placed on a heijunka board to signal that another batch of that resin type should be made at the appropriate time within the production-leveled schedule.

Another example involves a two-bin system, where the bins themselves act as the kanbans. Plastic pellets of ten different types are manufactured and placed in bins. The bins feed a downstream press operation that makes high strength, high temperature elastomeric O-rings. There is a storage rack (Figure 14.13) with two slots for bins of each pellet type. When a specific type—say, Type 321—is needed at a press, the bin is removed and taken to the press. If there is material left in the bin, the bin is returned to the rack. Once the bin is empty, it is returned to the pellet manufacturing operation as a signal to make another bin of Type 321. Each bin is sized to hold enough material to satisfy average demand for that type during the lead time required to fill the other bin. Because consumption of that type may be higher than average during that lead time, and because it may take somewhat longer than expected to refill a bin, each bin is sized to also hold some safety stock.

Bin size = Average consumption during Average Lead Time + Safety Stock

Thus, the bins act as the signal, or kanban, giving permission to produce more of that material type.

To summarize, pull systems in process plants use a variety of kanban and ConWIP flow control techniques, and traditional cards and totes as well as takt boards and electronic displays to convey production needs visually.

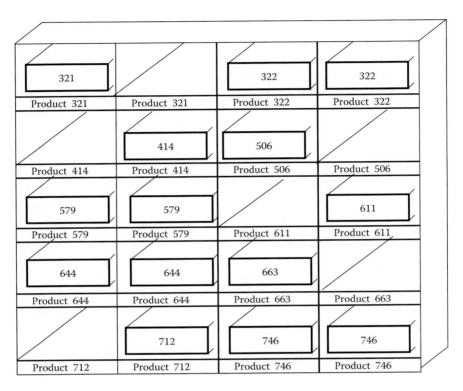

FIGURE 14.13
Pull using a two-bin kanban approach.

PUSH IN REAL LIFE

A dramatic example of the problems that a push system can cause occurred in California early in 2005. A process was being operated in push mode, a constraint was encountered, and as a result, eleven people lost their lives.

The background on the event is as follows. A 26-year-old California man, Juan Manuel Alvarez, reportedly intended to commit suicide and drove his Jeep Grand Cherokee onto a local railroad track. After sitting parked on the track for some time, he left the Jeep and walked away. A Metrolink commuter train hit the Cherokee, causing passenger cars to jackknife and derail. Eleven people were killed by the crash and almost 200 others were injured. (It was initially believed that Alvarez had second thoughts and decided not to commit suicide. Police later believed that he never intended suicide, but wanted to do something dramatic to gain attention. Murder-with-intent charges were filed against him.)

It may be hard to fathom how a two-ton vehicle could cause a train weighing hundreds or thousands of tons to derail. The answer is that the train was being pushed. California is one of a few states that allow trains to operate in what is called "push–pull configuration," where the train is pushed by a locomotive at the rear when traveling in one direction, and then pulled by the locomotive when traveling in the opposite direction. An investigation into the crash, and into railway safety in California in general, revealed that in the prior twenty-three-year period, thirteen passengers had been killed on trains being pushed, while none on trains being pulled. The special committee conducting the study recommended that California phase out the push–pull configuration.

The reason for mentioning this incident is that it forms a powerful visual analogy for the problems that can occur in a manufacturing process being operated in push mode. When everything is happening as expected, and no unexpected constraints are encountered, everything flows smoothly. However, when a constraint arises, whether it be a slowdown in customer demand, an upset in the manufacturing process, or a Jeep Cherokee, the motive force in a push system continues to move more and more material toward the constraint in accordance with the forecast, the MRP system, or whatever is driving the push signal. Although the results are not nearly as severe as in the train wreck, the jackknifed cars can be considered a metaphor for the inventory that can build up in a manufacturing process operated in push mode.

SUMMARY

Various books on lean offer differing and often contradictory definitions of what constitutes pull, so you may get confused trying to sort it out. The important thing is to understand why pull was developed and focus on the reasons that make pull a better replenishment system than a push system. Viewed from that perspective, a pull system is one in which day-to-day production is based on current conditions on the plant floor (rather than scheduled from a forecast), production is synchronized to true customer demand, and inventory is kept to the minimum needed for smooth flow.

By that definition, pull can be used to produce material needed to fill current customer orders as well as to replenish material that has been consumed by downstream processes.

The traditional way to manage pull in parts manufacture and assembly is through the use of kanban signals in the form of cards, containers, and available space in a storage area. Although kanban is sometimes used in process plants, the high number of product types often favors an alternate system called ConWIP (also called part-generic kanban), versus the more traditional part-specific kanban.

Pull systems inherently require that manufacturing equipment be started and stopped in accordance with kanban or ConWIP signals. With some process equipment, stopping in this way is not practical, because a shutdown and restart can cost hundreds of thousands of dollars. Consequently, many process lines make use of a push–pull interface, where material is pushed through the difficult-to-stop equipment to an in-process inventory. Material is then pulled from that inventory as needed to feed downstream operations.

Although pull can offer great benefit to process lines as well as assembly lines, you should not begin to design pull until the process has been stabilized and variation has been reduced as much as possible. If virtual cells are appropriate, they should be implemented before pull, as should product wheels. With all the foundational improvements successfully in place, you are then well prepared for a very successful and effective pull implementation.

15

Supermarket Design

The term "supermarket" refers to inventory that is managed in a way that facilitates a pull replenishment system. The basic concept is similar to that behind the typical food supermarket. There are storage locations (analogous to grocery shelf spaces) for each specific product. As material is needed by customers, either external customers or downstream steps in the manufacturing process, it is removed from the supermarket in the quantity required by the customer. The empty space provides permission to restock the space with the quantity of material that has been pulled. Restocking with more material than had been removed is not allowed: in a food supermarket, there is no shelf space available; in a manufacturing supermarket, the control mechanisms must also prevent producing more than was pulled.

With a properly managed supermarket, the pull system goals of preventing overproduction and of synchronizing flow to customer demand are being met.

It is not coincidental that the primary method for managing inventory in a pull environment would resemble that used in grocery supermarkets. Indeed, Ohno conceived the basic kanban and pull techniques from his analysis of American supermarkets done in the late 1940s. He realized that these methods provided a way to achieve his goal of just-in-time production.

The shelf space allocated to each product is based on the expected demand between restocking. Thus, if restocking is done frequently, less shelf space is required for each product. Working toward the lean goal of smaller lot sizes and more frequent production will reduce the quantity of each product needed in the supermarket, as well as the space needed to store it.

UNDERSTANDING THE SUPERMARKET CONCEPT

The rules for managing a supermarket are simple and straightforward:

1. Manufacturing is permitted to produce only the amount that has been removed from the supermarket.
2. Material consumption must be signaled by simple, visual means.
3. The material to replenish the supermarket must not include any defective material.
4. There should be an ongoing process to reduce the amount of each material needed in the supermarket.

Although the rules are simple to state, they can be difficult to follow. The third rule, about preventing defective materials from entering the supermarket, can be especially difficult in most process industry operations, where defects are not readily apparent and testing is required to identify them. Because this testing often takes time, the material can arrive at the supermarket before any defects have been identified, in violation of the third rule.

When designing supermarkets, one should note that the conceptual design of the supermarket and of its control signals and visual displays is usually done on an aggregate basis, but when determining supermarket quantities (how much shelf space is needed), it must be done on a specific SKU-by-SKU basis.

Designing and operating supermarkets is a matter of answering three questions:

1. When do we need to replenish a material in the supermarket?
2. How much of that material should be replenished?
3. How should the need to replenish be communicated to the producing step?

Then, you put processes in place to determine the answers to the first two questions on an ongoing basis, and you implement the answer to the third question.

Questions 1 and 2 are essentially inventory management questions. The subject of inventory management is complex, because it is multidimensional:

- Inventory can consist of cycle stock, safety stock, and defective or problem stock.
- Cycle stock can be based on demand history or on forecasts of future demand.
- Safety stock may be needed to protect against demand variability, forecast error, lead time variability, or production variability.
- Inventory can be replenished on a fixed interval or on a fixed quantity basis.
- Inventory can be replenished on a push or on a pull basis.

All these are explained in this chapter.

INVENTORY TYPES AND SUPERMARKETS

Basically, inventory exists anywhere in a process where flow is not continuous, where material stops moving for any reason for any length of time. The longer the time the material is stopped, the greater is the resulting inventory.

This includes raw material inventory, where raw materials are purchased in lot sizes greater than can be immediately consumed, which is almost always the case. The exception is a case where chemical ingredients are piped in, and flow can be turned on and off as needed.

This includes finished goods waiting for shipment to customers, in an MTS process.

This includes WIP being held to satisfy demand for specific materials not currently being produced by a step that produces several product types, such as one scheduled by product wheel. It also includes WIP temporarily held because of rate differences between adjacent steps in the process. When adjacent steps run at different rates, some temporary WIP is required to buffer the rate discontinuity. For example, sheet forming can generate a roll every fifteen minutes, while the bonder takes seventeen minutes to process a roll. Thus, bonding will not always be available to receive the roll coming off of the forming machine, so it must be buffered temporarily.

Transportation between steps in the supply chain also creates inventory, even though the material is moving. This is called transportation inventory or pipeline inventory, and is proportional to the flow rate and the distance traveled.

Supermarkets are needed only where inventory is intentionally being maintained to cover the cyclic nature of replenishment, which includes most raw material inventories, MTS finished goods inventories, and WIP to cover progression through a multiproduct cycle. In the other cases, temporary WIP buffers and pipeline inventory, the material should be allowed to flow as fast as it can, generally in a FIFO fashion.

INVENTORY COMPONENTS DEFINED: CYCLE STOCK AND SAFETY STOCK

Where inventory is deliberately being maintained, it generally has two components, cycle stock and safety stock. *Cycle stock* is the amount of a specific product to be made during the production cycle, to satisfy demand over the full cycle including the portion of the cycle when other products are utilizing the asset. For example, if the production process is based on a seven-day product wheel, the cycle stock for Material A would be seven days. If Material A occupies one day on the wheel, at the end of its production day there must be six days of material in the finished goods warehouse, or in downstream process steps and headed for the warehouse. That material is needed to satisfy demand for product A in the six-day interim until Material A will be made again. So the cycle stock for Material A includes the one day that was consumed while Material A was being produced, and the six days to satisfy demand during the rest of the cycle.

The second component of inventory is *safety stock*, material held to satisfy demand in cases where actual demand is higher than expected, or where the next cycle was late in starting.

Figure 15.1 shows a profile of inventory versus time for a single SKU in a case where cycle stock and safety stock are present. In production period P1, cycle stock is produced, to raise the level to A. Demand during the next cycle, D1, is equal to the average demand, so the cycle stock is consumed, but safety stock is not. Production P2 raises total inventory back to level A. Demand during the next cycle, D2, is higher than average, so that in addition to consuming all the cycle stock, some of the safety stock is needed. This would also be the case if it took longer than average for the process to complete its cycle and return to making this material. Thus, the safety stock will protect flow against either variation in demand or variation in

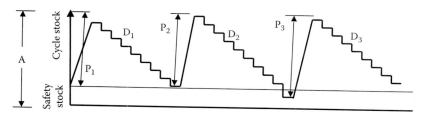

FIGURE 15.1
Cycle stock and safety stock.

supply lead time. Production P3 must be greater than average, to replace cycle stock plus the amount of safety stock that was consumed.

Cycle stock is based on the average demand expected. This can be based on either demand history or on a forecast. If previous demand is considered to be the best predictor of future demand, demand history should be used to set cycle stock. If there is a forecast that is believed to be a more accurate indication of future demand, cycle stock should be based on the forecast. Because forecasts can vary period by period, the cycle stock may be adjusted upward or downward each period in accordance with the forecast.

Some may consider that carrying safety stock is counter to lean, that it is waste, that it is material being stocked "just in case." Although it is waste, it is necessary to protect customers against the variations in our process and the variation the customers themselves present us with. Until these variations can be eliminated, smooth flow of products to customers requires safety stock. Even Henry Ford, who was relentless in driving out waste, recognized this when he wrote, "it is a waste to carry so small a stock of material that an accident will tie up production."

We have defined safety stock to be that required to protect against variation in both supply and demand. As noted in Chapter 12, some authors separate the two, while others combine them. If variation in supply and variation in demand are statistically independent, it is beneficial to combine them. Because periods of higher than average demand are statistically less likely to coincide with periods of long lead times, safety stock calculations based on the combined variance will give a lower inventory requirement than separate calculations based on each variance.

Safety stock is being carried because we intend to use it, and will use it frequently. This is in contrast with the guidance on safety stock usage given in some references, which is that safety stock be used only in extreme

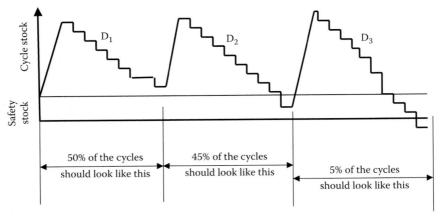

If inventory has been designed for a 95% service level....

FIGURE 15.2
Expect stockouts consistent with service level settings.

situations and only with management permission. It is important for you to understand that safety stock is there to cover normal, random variation in demand and/or supply. Because demand will be higher than average about half the time and lower than average the other half, one should expect to consume some of the safety stock during about half of the cycles, as depicted in Figure 15.2. Further, if safety stock is calculated based on a 95 percent cycle service level, stockouts should be expected on 5 percent of the cycles.

In addition to cycle stock and safety stock, supermarkets might also contain unintentional inventory, which is not planned and is certainly not desirable, but is the result of some dysfunction. This includes:

- Defective inventory, out of spec inventory
- Damaged inventory
- Obsolete inventory
- Slow moving inventory (due to poor demand management processes or forecasting error)
- Inventory resulting from product development test production

It is important to recognize when these forms of inventory exist, so that they are not counted as part of cycle stock or safety stock. If some of the inventory in a supermarket is of this type, and is not identified as such, the likelihood of stockouts will increase.

CALCULATING CYCLE STOCK

Cycle stock can be replenished on a fixed interval or on a fixed quantity basis. (These are the two major models; there are others, which will not be discussed here.) As the name implies, *fixed interval replenishments* occur on a regularly repeating cycle, where the time between replenishments may vary only slightly, but the quantity can vary significantly, depending on how much material has been consumed during the most recent cycle. *Fixed quantity replenishment* behaves the opposite way: The quantity is determined based on some specific criteria and doesn't vary. The interval can vary significantly, again based on the rate of consumption since the last replenishment.

Supermarkets can be refilled by either of these models; which one is most appropriate depends on the production process or the raw material ordering process.

Calculating Cycle Stock: Fixed Interval Replenishment Model

Figure 15.3 shows the inventory profile for a single material in a fixed interval strategy. The specific case shown is a product wheel, with a fourteen-day wheel time, but it could also depict raw material inventory for a material ordered every fourteen days. In a product wheel situation, we must make enough to last until the next production of this material, or

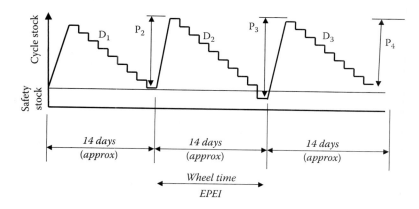

FIGURE 15.3
Inventory profile with fixed interval replenishment.

fourteen days worth. So the cycle stock will be the average demand during a fourteen-day period, and the peak inventory will be cycle stock plus safety stock, or fourteen days worth plus safety stock. What actually gets produced when that spoke of the wheel comes around is not always the cycle stock, but depends on current inventory. In Period 2, for example, demand D2 is slightly greater than average, so some of the safety stock has been consumed. Thus, the quantity to be produced, P3, will include the normal cycle stock plus the amount of safety stock that was consumed.

The standard equations governing this model are as follows:

$$\text{Peak Inventory} = \text{Cycle Stock} + \text{Safety Stock}$$

$$\text{Average Inventory} = \tfrac{1}{2} (\text{Cycle Stock}) + \text{Safety Stock}$$

These equations are accurate for purchased materials, that is, for materials received as a complete lot, equal to the cycle stock. They are approximations when applied to materials being produced within our process, because some of the cycle stock is being consumed by downstream steps during the production cycle. This has a minor effect on products that occupy a small portion of the production cycle, but can be significant if a product occupies a large portion of the cycle.

The following equations apply to those situations:

$$\text{Peak Inventory} = \text{Cycle Stock} \left(1 - \frac{D}{PR} \right) + \text{Safety Stock}$$

$$\text{Average Inventory} = \tfrac{1}{2} \left(\text{Cycle Stock} \right) \left(1 - \frac{D}{PR} \right) + \text{Safety Stock}$$

where D is the demand for that material per unit of time, and PR is the production rate, the total quantity produced over that same time.

The quantity to be produced on any cycle will be:

$$\text{Quantity Produced} = \text{Cycle Stock} + \text{Safety Stock} - \text{Current Inventory}$$

Because the current inventory will on average be approximately equal to the safety stock, the quantity produced will generally be approximately the cycle stock.

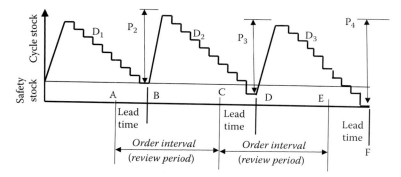

FIGURE 15.4
Fixed interval replenishment with lead time.

If this model is used to order raw materials, there will generally be a lead time before the material is received, so the profile will look like Figure 15.4. In this case, when the normal order interval begins, at point A, for example, enough material must be ordered to cover demand during the lead time as well as that needed to restore total inventory to the cycle stock plus safety stock target. Thus, the amount to be ordered at point A is:

Order Quantity = DDLT + Cycle Stock + Safety Stock − Current Inventory

where DDLT = demand during lead time.

The current inventory will typically be approximately DDLT plus safety stock, so the amount ordered will be approximately the cycle stock.

If the demand during the lead time is greater than average, as shown in the lead time C–D, safety stock will prevent a stockout, but when the new order arrives, the order quantity will not bring total inventory up to the cycle stock + safety stock target. If safety stock has been calculated appropriately, that shortfall will be covered.

This method of replenishment is sometimes referred to as a fixed order interval (FOI) model or a periodic review system. This is the same replenishment process as is used in a grocery supermarket where the shelves are restocked on some regular basis, say every Friday morning. The interval is fixed, Friday to Friday, but the quantity will vary based on the amount customers have pulled from the shelf since the previous Friday.

Calculating Cycle Stock: Fixed Quantity Replenishment Model

Fixed quantity replenishment is an alternative to fixed interval, and is used when there is some benefit in buying or producing materials in specific quantities. In the process industries, some materials are received in tank trucks, so transportation economics suggest buying in truck quantities. Suppliers of cardboard packaging materials are quite willing to print the customer's name, logo, and other information on the stock, but only if a certain minimum quantity is ordered, so it generally is advantageous to order that quantity. In other cases, an economic order quantity (EOQ— see Chapter 12) calculation, which balances ordering costs with inventory carrying costs, will be used to optimize order quantity.

In the production process, there is often a specific campaign size that best balances changeover cost with inventory carrying cost, as determined by an EOQ calculation. That quantity would be used to replenish finished product inventory on a fixed quantity basis.

An inventory profile for a single material replenished using a fixed quantity model is illustrated in Figure 15.5. Because the order quantity Q is already known, the question to be answered in this case is when it is time to place the next order. Whenever current inventory falls to or below the order point, a new order is placed. The time between orders can be variable: Order interval B–C is slightly shorter than interval A–B, because demand D3 is greater than demand D2. Thus, in contrast with the fixed

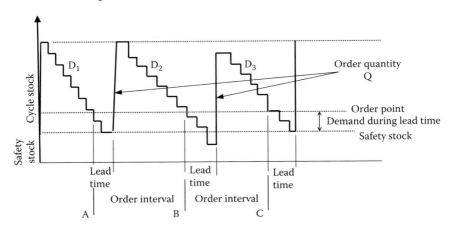

FIGURE 15.5
Inventory profile with fixed quantity replenishment.

interval model, the interval here will vary, while the quantity ordered remains fixed. In the simplest case, the order point is calculated as:

$$\text{Order Point} = \text{DDLT} + \text{Safety Stock}$$

where DDLT = demand during lead time.

In a perfect world, with no safety stock, the new order would arrive just before a stockout would occur. However, in the real world, the new order may arrive late or the demand during the lead time might be greater than average, so we need safety stock to cover those situations. Thus, the order point is set so that on average, the new order will arrive just as inventory falls to the safety stock level.

This model is often called Continuous Review, because it assumes that the inventory level is being continuously monitored, and that the new order is placed immediately when the inventory falls to the reorder point. In many cases that is true, but in some situations inventory is not being checked continuously. Inventory status may be checked once per day, per week, or at some other frequency. The time between inventory examinations is called a review period. If the replenishment process includes a review period, it must be accommodated in the order point. In these cases the order point calculation is:

$$\text{Order Point} = \text{DDRP} + \text{DDLT} + \text{Safety Stock}$$

where DDRP = demand during review period.

In cases where the lead time is very long, this equation will result in a very large order point, which is shocking to some. However, in either of these situations, current inventory includes not only the inventory on hand, but also inventory currently in transit and orders placed but not filled.

So an order actually gets placed when:

$$(\text{Inv on Hand} + \text{Inv in Transit} + \text{Unfilled Orders}) < $$
$$(\text{DDRP} + \text{DDLT} + \text{Safety Stock})$$

In this replenishment model, the cycle stock is the order quantity Q, and as in the fixed interval model:

$$\text{Peak Inventory} = \text{Cycle Stock} + \text{Safety Stock}$$

$$\text{Average Inventory} = \tfrac{1}{2}(\text{Cycle Stock}) + \text{Safety Stock}$$

$$\text{Cycle Stock} = Q$$

As before with the fixed interval model, these equations must be adjusted in cases where a substantial portion of the cycle stock is consumed during its production:

$$\text{Peak Inventory} = \text{Cycle Stock}\left(1 - \frac{D}{PR}\right) + \text{Safety Stock}$$

$$\text{Average Inventory} = \tfrac{1}{2}\left(\text{Cycle Stock}\right)\left(1 - \frac{D}{PR}\right) + \text{Safety Stock}$$

This fixed quantity model is also known as a continuous review model, an ROP (re order point) model or a *Q,r* model (where *r* is the reorder point).

An important attribute of a fixed quantity, continuous review model is that it will always require less safety stock than a fixed interval model, for the same degree of variability and desired customer service. The former needs safety stock protection only during the lead time, where the latter requires safety stock protection during lead time and the interval duration. If, however, the fixed quantity model is not continuous review, but has a review period, then additional protection will be needed during the review period. Thus the safety stock advantage diminishes as the fixed quantity review period approaches the fixed interval duration.

Because the fixed quantity model usually requires lower inventory than the fixed interval model, it is often used with materials of relatively high value. It is also used where there is a strong economic reason to buy, produce, or ship in specific quantities. Information systems must be in place to support continuous or very frequent review of current inventory levels for a fixed quantity process to provide full advantage. If it is very difficult or costly to get frequent inventory level updates, then a fixed interval process may be preferable. A fixed interval model may also be chosen because its structure and predictability enables better planning and scheduling of support activities like preventative maintenance tasks and QC lab tests.

CALCULATING SAFETY STOCK

Safety stock is inventory carried to prevent, or reduce the frequency of, stockouts, and thus provide better service to customers. Safety stock can be used to accommodate:

- Variability in customer demand or in demand from downstream process steps (where demand history is used to set cycle stock or order points)
- Forecast errors (where forecasts are used to set cycle stock targets or order points)
- Variability in supply lead times
- Variability in supply quantity

If these variabilities are random, and are reasonably normally distributed, the following calculations will result in appropriate safety stock levels. If not, they may still give some guidance, and are generally preferable to the sometimes recommended rules of thumb, that safety stock be set at 10 percent, or 20 percent, or 50 percent, of cycle stock.

Variability in Demand

If variability in demand can be expressed as a standard deviation (σ_D), safety stock can be calculated by:

$$\text{Safety Stock} = Z \times \sigma_D$$

where Z is the number of standard deviations of demand variation we wish to cover. If Z is zero, if no safety stock is carried, there will still be enough inventory on half of the cycles. Demand will be less than average half of the time, so all demand will be met even with no safety stock. If Z is 1.0, the safety stock will protect against one standard deviation, so that there will be enough inventory 68 percent of the time. The percentage of cycles during which the safety stock should prevent a stockout is called *cycle service level* (CSL). Figure 15.6 shows the relationship between Z and CSL. As can be seen, the relationship is highly nonlinear: Higher CSL values (that is, lower potential for stockout) require disproportionally higher safety stock levels. Statistically, 100 percent CSL is impossible.

Desired Cycle Service Level	Z Factor
84	1
85	1.04
90	1.28
95	1.65
97	1.88
98	2.05
99	2.33
99.9	3.09

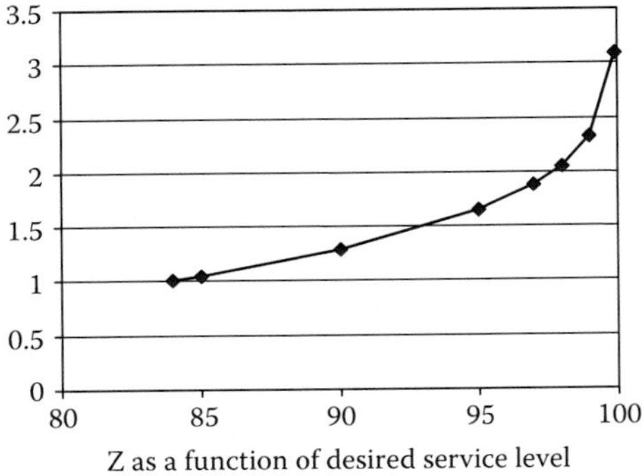

Z as a function of desired service level

FIGURE 15.6
Relationship between service factor and service level.

Typical service level goals are in the 90 to 98 percent range, but good inventory management practice suggests that rather than using a fixed Z value for all products, Z be set independently for groups of products based on strategic importance, profit margin, dollar volume, or some other criteria. Doing this will place more safety stock in those SKUs with greater value to the business, and less safety stock in the products believed to be less important to business success. More is said about this in Chapter 16.

The safety stock equation assumes that the standard deviation of demand is calculated from a data set where the demand periods are equal to the total lead time, including any review period. If not, an adjustment must be made to the standard deviation value to statistically estimate what the standard deviation would be if calculated based on the periods equal to the lead time. As an example, if the standard deviation of demand is calculated from weekly demand data, and the total lead time including the review period is three weeks, the safety stock must cover demand variability over three-week periods. The standard deviation of demand calculated from a data set covering three-week periods would be the weekly standard deviation times the square root of the ratio of the time units, or the square root of 3. Bowersox and Closs, in *Logistical Management*, use the term *performance cycle* (PC) to denote the total lead time. If we let $T1$ represent the time increments from which the standard deviation was calculated (one week in this example), then

$$\text{Safety Stock} = Z \times \sqrt{\frac{PC}{T1}} \; \sigma_D$$

When procuring raw materials, the performance cycle includes the time to:

- Decide what to order (order interval or review period)
- Communicate the order to the supplier
- Manufacture or process the material
- Deliver the material
- Perform a store-in

Inside our own manufacturing facility, the performance cycle includes the time to:

- Decide what to produce
- Manufacture the material
- Release the material to the downstream inventory
- Return to the next cycle
- If we are carrying inventory in a finished product warehouse, and customers allow a delivery lead time greater than the time needed to deliver to the customer, then the remaining customer lead time can be subtracted from the Performance Cycle

If cycle stock has been calculated from historical demand, the variance used in the safety stock calculation should be based on past demand variation. If forecasts are used to set cycle stock, the parameter necessitating protection is forecast error. Standard deviation of forecast error would replace standard deviation of past demand in the safety stock formula, which would become:

$$\text{Safety Stock} = Z \times \sqrt{\frac{PC}{T1}} \times \sigma_{Fcst\ Err}$$

If there is bias in the forecast, it must be removed for the safety stock calculation to be valid. (Dealing with forecast bias is beyond the scope of this book.)

Variability in Lead Time

The equations in the preceding section calculate the safety stock needed to mitigate variability in demand or forecast error. If variability in lead time is of concern, the safety stock needed to cover that is:

$$\text{Safety Stock} = Z \times \sigma_{LT} \times D_{avg}$$

The average demand term (D_{avg}) is in the equation to convert standard deviation of lead time expressed in *time units* into production *volume units* (such as gallons, pounds, rolls).

Combined Variability

If both demand variability and lead time variability affect a given super-market inventory, the safety stock required to protect against each can be combined statistically, to give a lower total safety stock than the sum of the two individual calculations. If demand variability and lead time variability are independent, that is, the factors causing demand variability are not the same factors influencing lead time variability, and if both variabilities are reasonably normally distributed, the combined safety stock is Z times the square root of the sum of the squares of the individual variabilities:

$$\text{Safety Stock} = Z \times \sqrt{\frac{PC}{T1}\sigma_D^2 + \sigma_{LT}^2 D_{avg}^2}$$

If σ_D and σ_{LT} are not statistically independent of each other, this equation can't be used, and the combined safety stock is the sum of the two individual calculations.

$$\text{Safety Stock} = \left(Z \times \sqrt{\left(\frac{PC}{T1}\right)} \times \sigma_D \right) + \left(Z \times \sigma_{LT} \times D_{avg} \right)$$

Cycle Service Level and Fill Rate

The equations in the preceding sections will predict the safety stock needed so that a certain percentage, say 95 percent, of the replenishment cycles will be completed without a stockout. This is often called cycle service level. Business leaders often want to control the percentage of total volume ordered that is available to satisfy customer demand, not the percentage of cycles without a stockout. The former quantity is called fill rate, and is often considered to be a better measure of inventory performance. Figure 15.7

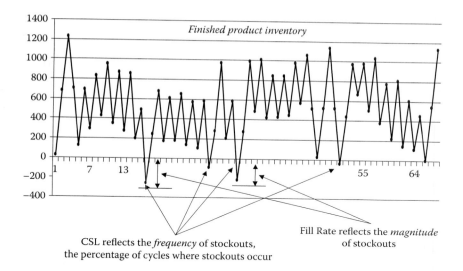

CSL reflects the *frequency* of stockouts,
the percentage of cycles where stockouts occur

Fill Rate reflects the *magnitude*
of stockouts

FIGURE 15.7
Cycle service level and fill rate.

illustrates the difference. Where cycle service level is an indication of the frequency of stockouts, without regard to the total magnitude, fill rate is a measure of inventory performance on a volumetric basis.

The specific calculations of safety stock required to achieve a desired fill rate are beyond the scope of this book. An excellent discussion can be found in Chopra and Meindl's *Supply Chain Management.* However, some observations are in order. With stable demand patterns and supply behavior (that is, low standard deviations of demand and lead time) fill rate will generally be higher than cycle service level, as illustrated in Figure 15.8. Although stockouts will occur, with low supply and demand variability the magnitude of each stockout will be small. With high variability in either demand or lead time, or both, the opposite will usually be found. Figure 15.9 illustrates a case where demand variability is high, where the standard deviation of demand is half of the average demand. Although there are few stockouts (because of the safety stock being carried) the magnitude of any stockout can be quite high. Thus, in this case, the fill rate is actually less than the CSL.

Although this chapter has described the fundamentals of safety stock determination in the most common situations, more thorough explanations can be found in Chopra and Meindl's *Supply Chain Management* and Bowersox and Closs's *Logistical Management.*

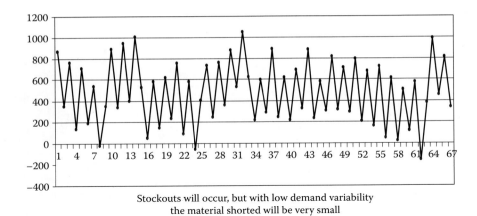

Stockouts will occur, but with low demand variability
the material shorted will be very small

FIGURE 15.8
Inventory profile with low demand variability (CV = 0.2).

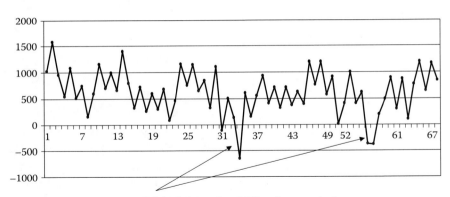

With high demand variability, there may be few
stockouts, but the amount can be quite large!

FIGURE 15.9
Inventory profile with high demand variability (CV = 0.5).

EXAMPLE: THE PRODUCT WHEEL
FOR FORMING MACHINE 1

As an example of the use of these calculations, consider the inventory levels required in the supermarket being supplied by virtual Cells 1 and 2, consisting of Forming Machine 1 and Bonder 1 and Forming Machine 2 and Bonder 2, respectively, shown in Figure 15.10. As Slitter 1 pulls a roll

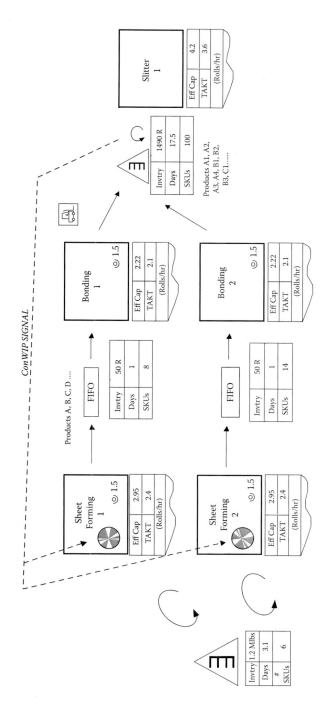

FIGURE 15.10
Paper Forming Cells 1 and 2.

from the supermarket, that creates a ConWIP signal that is loaded onto the schedule for the product wheel on either Forming Machine 1 or Forming Machine 2, depending on which cell is assigned that product. The loading of products needed by the supermarket onto a product wheel schedule is analogous to placing kanban cards on a heijunka board. When the appropriate spoke comes up on the forming product wheel, all material of that type gets formed, and then flows in FIFO fashion to the bonder. The bonder may have to go through temperature changeovers during a given forming spoke; a single product coming from Forming Machine 1 can be converted to several products in the bonding process. Thus, Product A coming from Forming Machine 1 will generate A1, A2, A3, and the other Product A derivatives, in the bonding process. This is the primary reason that material is held up in the FIFO buffer within this cell: It is waiting for bonding temperature to be changed to meet its specifications.

Looking specifically at the supermarket requirements for product A1, which is part of product family A made on Forming Machine 1:

- Weekly demand = 50 rolls
- Standard deviation of weekly demand = 10 rolls
- Standard deviation of lead time = 0

Because forming is running a product wheel, this replenishment will be on a fixed interval basis. With a seven-day wheel time, and product A being produced every cycle, the cycle stock for any product in the A family is seven days worth. So for product A1, the cycle stock is 50 rolls.

The lead time is stable and predictable; forming reliability is high enough that the wheel time never exceeds the seven days. Bonder reliability is high enough that the delay through the FIFO buffer never exceeds one day. So with a standard deviation of lead time of zero, the safety-stock requirement can be calculated from:

$$\text{Safety Stock} = Z \times \sqrt{\left(\frac{PC}{T1}\right)} \times \sigma_D$$

If 95 percent is the desired cycle service level (the business can tolerate stockouts of this product on no more than 5 percent of the replenishment cycles, slightly more than two per year), the Z value can be found in Figure 15.6 to be 1.65. PC, the performance cycle affecting the pre-slit supermarket, is the sum of the seven-day wheel time and the one-day delay through the bonder FIFO, for a total of eight days. T1, the time increments from which σ_D was calculated, is seven days.

Thus:

$$\text{Safety Stock} = 1.65 \times \sqrt{^8\!/_7} \times 10 \text{ rolls} = 18 \text{ rolls}$$

$$\text{Peak Inventory} = \text{Cycle Stock}\left(1 - \frac{D}{PR}\right) + \text{Safety Stock}$$

where D = 50 rolls/week and PR = 496 rolls/week:

$$\text{Peak Inventory} = 50 \text{ rolls} \left(1 - {}^{50}\!/_{496}\right) + 18 \text{ rolls} = 63 \text{ rolls}$$

$$\text{Average Inventory} = \frac{1}{2}\left(\text{Cycle Stock}\right)\left(1 - \frac{D}{PR}\right) + \text{Safety Stock}$$

$$\text{Average Inventory} = \frac{1}{2}\left(50 \text{ rolls}\right)\left(1 - {}^{50}\!/_{496}\right) + 18 \text{ rolls} = 41 \text{ rolls}$$

When the spoke for product A is being run on the product wheel, it should make enough so that A1 can be brought back up to the 63-roll level by the end of its production. This will require production of 68 rolls minus the current inventory, because 5 rolls will be consumed while it is being produced. Of course, Forming Machine 1 also needs to produce enough of Material A to bring A2, A3, A4, and the rest of the A family back to their respective supermarket levels.

In this process, the amount to be produced on any spoke is determined at the start of that spoke, so the roll consumption inventory quantity is up to date. In some cases, the entire wheel production is set at the start of the wheel cycle, so the exact consumption before the start of any spoke must be predicted. This adds a review period, described earlier, which must be included in the performance cycle used in safety-stock calculations.

If there were variability in lead time, if forming equipment failures caused the seven-day wheel time to vary, more safety stock would be required to meet the inventory performance goals. If, for example, wheel time varied with a standard deviation of ½ day, or 0.07 weeks, the safety stock calculation would be:

$$\text{Safety Stock} = Z \times \sqrt{\frac{PC}{T1}\sigma_D^2 + \sigma_{LT}^2 D_{avg}^2}$$

$$\text{Safety Stock} = 1.65 \times \sqrt{(\text{\textonesuperior}\!/\!_7)10^2 + (0.07)^2 50^2}$$

$$\text{Safety Stock} = 1.65 \times \sqrt{114.3 + 12.2} = 19 \text{ rolls}$$

Two things can be seen from this result. The first is that in this example the demand variability has the dominant influence on safety stock requirements: Its effect on safety stock is almost ten times that of lead time variability. It is often the case that one factor or the other will dominate the calculation; it is important to recognize that, so that improvement efforts can be focused on the most appropriate things. In this case, if we decide to reduce the need for safety stock, it is far more productive to work on demand variability than on lead time variability. The second observation is that the influence of lead time variability is so small that safety stock requirements increase by only 1 roll, to 19 rolls, compared to what it was without considering lead time variation.

To summarize, if the supermarket for A1 includes 19 rolls of safety stock, stockouts should be expected to occur on no more than 5 percent of the cycles, or during two to three of the weekly product wheel cycles each year. The fill rate in this case would be 99.5 percent (see Chopra and Meindl's *Supply Chain Management* for one example of the calculations involved.), so that even though we expect two or three stockouts per year, we expect to satisfy 99.5 percent of the demand on a volume-of-rolls basis. So the two or three stockouts would comprise a total of 12 rolls, out of the 2,600 that were required to satisfy customer demand for products made from Product A1.

ALTERNATIVES TO SAFETY STOCK

These calculations sometimes result in safety stock recommendations that are more than the business leaders feel they can afford to carry. Be advised that there are alternatives. Sometimes, an expediting process can be designed that can prevent a stockout when safety stock is not sufficient to cover all random variation. For example, if the goal is 98 percent CSL, safety stock can be reduced by 38 percent (Z factor of 1.28 rather than 2.05) if calculated to give a 90 percent CSL where a contingency plan can be defined to prevent stockouts in the other 8 percent of cycles. The contingency plan must be planned and agreed upon in advance. It is unacceptable to ignore this step, hoping that something can be figured out when the time comes.

This practice is especially appropriate with very expensive products, which are very costly to carry in inventory. In one specific example involving an expensive but relatively lightweight product, total supply chain cost was reduced significantly by carrying small amounts of safety stock in overseas warehouses and then relying on air freight to cover demand peaks. The cost of air freighting a small percentage of total demand was minimal compared to the cost of carrying large safety stocks of this highly valuable material on an ongoing basis.

Another alternative to carrying safety stock is to consider if MTO or FTO is possible. If lead times allow it, MTO completely eliminates the need for any safety stock. If lead time commitments will not allow full MTO, FTO can locate the safety stock where it is generally far less differentiated, so that demand variability will be much less (on a relative basis) and safety stock requirements will be lower than they would be with finished product inventory. Customers will sometimes be willing to accept longer lead times for highly sporadic purchases, making FTO or MTO more of a possibility.

SIGNALING METHODS

There are a wide variety of methods that can be employed to communicate the need to produce material to replenish material pulled from a supermarket. Traditional kanban systems often use cards, which not only signal permission to produce, but also communicate the specific type to be made and the lot size represented by each card. Bins, totes, and carts can also provide that function and, therefore, are labeled with the same information. Empty space on a shelf (the prototypic supermarket model), an empty space on the floor, and an empty slot in a rack system are also used to signal a need to replenish material; as with cards and bins, there must be placards or signs to indicate type and lot size.

Equipment used in process plants is usually controlled by PLCs (programmable logic controllers) and DCSs (distributed control systems). In many cases, these control systems are also monitoring in-process inventory, such as tank levels and slots currently occupied in a storage system. So they can be programmed to generate the pull signals automatically and transmit them to the previous step to begin replenishment. In these situations, the computer-driven displays can provide the visual information to operators, mechanics, and first line supervisors so that everyone associated with the process knows what is being produced and why. In some cases the electronic information is transcribed onto a handwritten takt board, as was described in Chapter 8.

THE ROLE OF FORECASTING

If what is produced on a day-by-day basis is scheduled from a forecast, that is a push system. However, there is a valuable role for forecasting in a pull system. Any time that demand may vary in a predictable manner, forecasts can allow higher service levels to be achieved with lower average inventory levels.

If, for example, there is a high degree of seasonality in customer demand (e.g., suntan lotion, fertilizers, exterior house paint, and barbeque sauce) and it is treated as normal, random demand variation, the safety stock required to cover it will be high. Even so, if calculated by traditional statistical methods, the safety stock will not be enough to cover demand during the peak season. So there will be excess inventory in the off-season, and not nearly enough in the peak season.

If, on the other hand, this seasonality is recognized and can be predicted with a reasonable degree of accuracy, the cycle stock targets can be adjusted up and down in accordance with the forecast, so that the amount of demand variability that must be covered by safety stock will be far less, and the net safety stock level will be less. In this mode of operation, the forecast is being used to set the inventory targets, and actual production is governed by current inventory levels versus the target.

The same principle applies to any predictable variation in demand, whether it is due to seasonal factors or to expected market trends. For example, if fuel prices continue their dramatic upward trend, the demand for alternate modes of transportation might be expected to increase. A company making bicycle tires might expect demand to rise, and adjust its cycle stock targets accordingly. Thus, although the inventory targets are adjusted based on forecasts, what is produced on a day-by-day basis is scheduled based on current inventory status.

If this use of a forecast seems more like push than pull, consider that Toyota expends considerable effort in forecasting, to know how many kanbans will be needed for each part type in each supermarket.

The important thing to understand is that inventory planning and target setting should be based on a forecast, but that actual production should be based on the current status of operations on the plant floor.

SUMMARY

Supermarkets are inventories managed in a way that follows pull replenishment principles; that is, material is put into the supermarket only to replace material that has been pulled out by the customer or downstream process step. Supermarkets exist only in an MTS environment; they are generally not needed with an MTO strategy, or beyond the FTO point in an FTO operation.

Although fundamental decisions on where to locate supermarkets and the strategy for managing them should be made on a product family basis, they must be designed and managed on an SKU-by-SKU basis.

The inventory for any SKU in a supermarket consists of cycle stock and safety stock. Cycle stock is determined by the length of the production campaign or by the order frequency in a process where replenishment is done at some fixed intervals. The cycle stock is the quantity ordered or the quantity produced in processes where those quantities are based on economic lot sizing criteria, such as EOQ calculations. Safety stock is there to protect against stockouts in the face of variable supply, production, or demand. It is determined by the variables we want to protect against, which may include any of the following: variability in lead time for purchased materials; variability in your production process when replenishing finished product inventory; variability in customer demand; forecast error; and combinations of these factors. A key decision in setting safety stocks is the level of protection desired; that is, how high you want the probability of not having a stockout to be. The higher the service level desired, the higher the safety stock requirements.

An important component of supermarket design is choosing the most appropriate signaling methods, where everyone involved can see in real time what must be produced to refill the supermarket.

When designed and managed effectively, supermarkets can enable customer takt to be met while minimizing inventory to the minimum level required for smooth flow.

16

The Importance of Leadership and Robust Business Processes

It is important to have business ownership, guidance, and direction when designing, implementing, and executing lean production. Lean is far more likely to be successful if guided by appropriate business leadership and clear performance targets. Business leaders must create and communicate a strong, clear vision of what lean will look like in their operations. They must connect high-level corporate goals to specific operational goals and metrics. Then they must champion and enable the changes required, by assigning people and then allowing them to spend the time necessary to make the lean effort successful. They must engage, train, and empower the entire workforce.

Ohno describes the Toyota concept of employee empowerment well. He likened the environment they were creating at Toyota to the autonomic nerve system that enables muscular subsystems to respond to immediate needs without having to consult the brain: "At Toyota, we began to think about how to install an autonomic nervous system in our own rapidly growing business organization. In our production plant, an autonomic nerve means making judgments at the lowest possible level; for example, when to stop production, what sequence to follow in making parts, or when overtime is necessary to produce the required amount." According to Ohno, "These [decisions] can be made by factory workers themselves."

Lean has a much better chance of becoming part of the manufacturing culture if that level of empowerment can be achieved. Alignment forces must be in place, however, so that chaos (people doing whatever they choose) is not the result. The alignment forces include KPI (key performance indicators) aligned with business objectives, standard work, and training.

Execution and improvement must be guided by the factors important to business success; these factors are translated into KPIs. Standard work must be followed; empowerment means improving the standard work. Change in standard work must include the involvement of the people who do the work, communication of the change, and proper training before the change is made. The role of the leader then changes from directing work to be done to one of coaching others in a learning environment.

Once all these things are in place and become institutionalized, leadership has an ongoing role in spending time in the workplace in order to recognize and reward good performance and in removing the barriers to maintaining and improving performance.

BUSINESS PRACTICES AND TARGETS

Then there are specific business processes that must be well executed in order for operations to have appropriate targets against which to execute. Many of the necessary targets are usually considered to be owned by the business or by the supply chain side of operations, but these targets also have a profound influence on the performance of the manufacturing operation.

ABC Classification

Several operational targets will lead to better business results if the targets are set differentially based on some type of product segmentation. Product segmentation can be based on total volume sold, on total dollars of sales, on variable margin, on margin contribution, or on some parameter related to strategic importance of each product to the business. One common practice is called ABC segmentation or classification, and ranks all products based on total dollars of sales. When this is done, it is generally found that the results come close to following the Pareto principle, that 80 percent of total revenue results from 20 percent of the products in the lineup. Figure 16.1 shows the product segments and revenue ranges suggested by APICS when ABC classification is done. Classifying and then grouping products on this basis allows various targets to be weighted in favor of those of most financial importance to the business.

Product Classification	Percent of Products	Percent of Revenue Dollars
A	10%–20%	50%–70%
B	20%	20%
C	60%–70%	10%–30%

FIGURE 16.1
Guidelines for ABC classification.

If classification is done on some other basis than sales volume or sales revenue, the Pareto principle may not apply. It may be, for example, that a small percentage of the products in the portfolio are considered to be of high strategic importance and, therefore, deserving of higher service level and safety-stock targets. That doesn't make the concept any less useful; just because the classified groupings don't follow Pareto doesn't make the classification any less valuable to the business.

The key point is that dividing the total product portfolio into groups (in other words, segmenting it) allows customer service goals, inventory managing guidelines, safety-stock levels, and a variety of other measures to be set individually for each segment. This then provides operations with a basis on which to prioritize production and material management in alignment with business goals.

Customer Lead Times

The business should decide and specify the lead time commitments that will be made to customers when they place orders. The lead time commitments are usually short—a few days—for MTS products because the premise is that they will be in finished product inventory at the time the order is received. Longer times, ranging from several days to several weeks, are generally specified for MTO products, to allow time for them to be produced. FTO products may have somewhat shorter lead times than MTO products.

Lead times may also be differentiated based on ABC classification or on some other criteria. Business strategy may suggest, for example, that shorter lead times be offered for newly developed products at the expense of the older, more mature products.

Customer Service Levels

Service levels, whether calculated by cycle service level (CSL) or by fill rate as defined in Chapter 15, may be differentiated on any one of several criteria. The business may decide to provide higher performance levels for products judged to be strategically most important to business success, and lower service levels to products closer to commodity status. Alternatively, products typically bought by the most important customers might get service level priority over those typically sold to less important customers. Because these service level targets will determine the amount of safety stock to be carried for each product, this provides a mechanism for the business to place its inventory dollars where they are believed to have greatest value to the business.

MTS, MTO, and FTO

The decisions about which products are to be made to stock and which will be made to order are important, as they have an impact on both lead times that can be promised to customers and on inventory requirements. If all product are MTS, and the appropriate safety stock levels are carried, customers can be promised short lead times. However, that typically requires large inventories, which can be eliminated by an MTO strategy. In some cases, long manufacturing lead times will require an MTS strategy; in others, FTO or full MTO may be possible. In many cases, a mix of MTS and MTO will optimize costs and performance.

The criteria to be used for determining MTO and MTS products should be clearly understood and documented. Products should be tested against those criteria on some regular frequency.

Demand Variability Analysis

It is often helpful to analyze demand variability for each final product or SKU. Products can be ranked by increasing coefficient of variation (CV), a measure of relative variability:

$$CV = \frac{\sigma_D}{D_{avg}}$$

Those products with a high CV, say a value greater than 0.5, will require much higher safety stocks to achieve a desired service level than those with more moderate CVs. In these cases, the business leaders often decide to specify that all products above a certain CV value will be MTO, and that lead time promises made to customers will be set to allow enough time to make to order. In other cases, this policy is applied to C products only, products with low demand volume and high demand variability. Where the marketplace will tolerate it, it makes sense to make products with sporadic demand on an MTO basis rather than carrying inventory that may not be consumed until some undefined time in the future.

Classifying products on an ABC basis and then on CV within each ABC classification allows business leadership to decide what strategies are most appropriate for their various product segments.

Protection (Safety Stock or Contingency Processes)

Chapter 15 discussed the use of predefined expediting processes as a lower cost way (in some situations) to achieve desired service levels compared to carrying large safety stocks. These are economic decisions that the business must make with input from operations on feasibility and practicality questions. Any contingency processes must be thoroughly planned in advance and be able to be executed without disrupting normal manufacturing and supply chain flow. The additional costs involved in expediting, such as air freight, overtime, or external subcontracting, must be committed to in advance.

SKU Rationalization

Many businesses have targets to gain a certain percentage of total revenue from new products, for example, products introduced within the past year, or within the past three years. As new products are introduced, it is important to reevaluate the existing products to weed out the ones that no longer make economic sense to carry in the portfolio. If SKU count is allowed to grow without limit, manufacturing complexity, lead times, and inventories will grow while customer delivery performance may suffer. If a business has an active SKU containment or product line optimization process, production processes will be less complex and the performance targets easier to reach.

Integrated Business and Operations Planning

The policies, practices, and targets in the preceding sections are separate and distinct but are highly interrelated and interdependent. Therefore, they should be evaluated and determined in an integrated planning process. Because this must be done by the business leaders with significant input from operations, this ongoing evaluation of policies and targets is best done in a regularly occurring process involving both functions. None of these is a singular, one-time decision. All must be revisited and updated periodically, with the frequency depending on the dynamic nature of the business. In a kinetic business environment, policies and targets should be reexamined quarterly or even monthly. In a more stable environment, bi-annually or annually may be sufficient. The key is that they be reviewed at some regular interval, with input from all groups with relevant input.

POOR BUSINESS PRACTICES

There are business practices that will hamper manufacturing and supply chain performance, and form barriers to becoming lean. Unfortunately, they are seen too frequently in the process industries.

Dictating Low Safety Stock Levels

Some business leaders are surprised at the levels of safety stock the calculations recommend, and insist that none or only a fraction of the recommended amount be held, in order to maintain carrying costs within targets. They then have unrealistic expectations for service levels and are disappointed when they are not met.

Expecting Abnormally Short Lead Times

Business leaders sometimes agree on customer lead times that are realistic and attainable based on manufacturing and supply chain performance. But when a customer requests delivery within a shorter timeframe, they expect that customer to be accommodated. If the manufacturing operation tries to comply, it can be disruptive to production scheduling processes and to material flow, and undo all the good that lean flow principles have brought to the operation. If it is important to be able to accommodate short lead time requests, then the material management and manufacturing processes must be so designed. If management has expectations that differ from those on which the manufacturing systems were designed, attempts to meet those expectations will add waste and cost, and reduce effectiveness.

Expecting Perfect Customer Service

Some business managers expect 100 percent service levels even though the supply chain parameters weren't designed on that basis. They are unwilling to tolerate any stockouts or unaccepted orders. They may have agreed to a 95 percent service level because it sounded like a high number, but are disappointed if there is a stockout once every twenty cycles.

Reducing Inventory at Year End

Some businesses reduce inventory arbitrarily at year end to make accounting results look better. Unfortunately, this behavior is seen quite frequently. Although this may reduce taxes and make year-end metrics look better, it is extremely disruptive to smooth flow through manufacturing operations and through the entire supply chain. It can take weeks to fully recover. Unfortunately, the costs in lost sales, overtime, and expediting are often not tied back to the correct root cause, so the behavior persists.

Pulling Next Quarter's Sales Ahead

When a business is approaching the end of the accounting quarter and sales targets are not likely to be reached, there is sometimes the tendency to encourage customers to accept next quarter's shipments early. In some cases they are even offered incentives to do so. When this happens, the manufacturing operation may have to delay other production to meet the accelerated demand. This may require product wheel cycles to be interrupted with non-optimal transitions. In short, it can be extremely disruptive to operations and to supply chain performance, and add cost. It also distorts true demand patterns and creates apparent demand variability. And in the end, it creates no additional value; it just makes the current quarterly results look better at the expense of next quarter's results. And because the next quarter starts with some of its normal demand already satisfied, they have to play the game at the end of the next quarter just to fill the hole they created at the start. In an article in a recent issue of APICS Magazine, Ronald Althaus called this practice "silent seasonality."

Obsessing over Cost Reduction

Many businesses have aggressive cost reduction targets. Unfortunately, in pursuing them they remove necessary cost as well as unnecessary cost. When this happens, the performance and indeed the health and stability of the operations deteriorate over time. This is like removing the muscle as well as the fat from the operation. If the focus could be redirected from cost reduction to waste elimination, as lean teaches, the unnecessary cost would be removed but the necessary cost would remain. So the muscle, health, and stability of the operation would be intact.

If lean leaders in the operation see any of these dysfunctional practices, they have an obligation to try to educate the business leaders on the negative effects of these poor practices.

INAPPROPRIATE USE OF METRICS

Some traditional manufacturing metrics not only ignore or discount the benefits resulting from lean improvements, but may actually discourage lean behaviors and practices. The topic of accounting for lean is beyond the scope of this book, and is well covered in the literature. However, two of the more frequent offenders will be described briefly:

- Pull systems inherently require that production equipment be stopped whenever enough material has been produced to satisfy immediate demand. This typically causes productivity measures to fall, including fixed cost productivity and equipment productivity. Any measures that track volume of production as a percentage of some relatively fixed basis may be lowered by a well-functioning pull system. Managers held accountable for those metrics must be given the latitude to operate in a way that best meets overall business objectives versus performing to satisfy some arbitrary, sub-optimizing targets. Running the process when there is no pull signal just to achieve productivity targets makes as much sense as driving an expensive new sports car around the block for several hours each day just to reduce the cost per mile.

- If equipment is not fully utilized, if it is down for lack of demand for some portion of available time, it may be appropriate to utilize that time to change products more frequently, to run smaller production campaigns. This will cause more time to be spent in product changes, but that time is essentially free if the equipment is currently underutilized. Smaller, more frequent campaigns reduce inventory and are very much in alignment with lean principles. However, as product changeover time counts against the OEE or UPtime metric, and more time is now being spent in product changeovers, those measures will suffer even though a beneficial change is being made. Those using OEE metrics as a gauge of equipment performance must balance it against other operational metrics. A conflict may arise from the fact that operational responsibility is often stove piped, with different individuals or groups accountable for OEE metrics and for inventory metrics.

SUMMARY

The transition from traditional manufacturing practices to lean production must be owned, driven, and led by the business leadership to have any reasonable chance of being successful. Senior management must engage in the appropriate business processes, make sound decisions regarding manufacturing performance targets, and ensure that metrics are aligned with the higher level business needs. The guidelines and targets that the operation is expected to meet should be explicit, clear, and consistent, and should be the basis for ongoing dialog between business leadership and operations leaders.

The managing processes and factors necessary for lean success include:

- Segmentation of the portfolio so that resources can be focused on the products with greatest strategic value to the business
- Alignment on the lead time promises to be made to customers
- Agreement on acceptable stockout rates and service levels

- Determination of which products are to be MTO, MTS, and FTO based on customer expectations and on manufacturing capability
- A process for sunsetting obsolete products, both to reduce manufacturing complexity and to prevent creation of slow-moving and non-moving inventory.

All these are best done in an integrated planning/managing process including business, marketing, supply chain, and manufacturing leaders.

Part V

Appendices

Appendix A

Determination of Appropriate Raw Material Inventory

I mentioned in Chapter 5 that the initial raw material quantities in the sheet forming plant were much higher than needed. Referring to Figure 5.1, you can see that Supplier 1 delivers six types of material at a total rate of 1.8 railcars per day. Railcars are likely to be carrying only one material type per trip. The inventory of any one of these six materials will consist of two components, cycle stock and safety stock. Cycle stock arises from the fact that a full railcar of a material is received at one time, and then consumed prior to the next arrival. Thus, the cycle stock portion of the inventory follows a sawtooth-shaped pattern as shown in Figure A.1. Safety stock is needed to prevent running out of a raw material when either the railcar arrives later than expected or consumption of the material between arrivals is higher than expected. So the total inventory has the sawtooth wave riding on top of the safety stock level. Because safety stock is providing protection against normal random variation in arrivals and in consumption, you should expect to dip into safety stock on approximately half the cycles. About half the time, you will have consumed all the cycle stock and a portion of the safety stock prior to the next arrival of that material, as happens in demand period D2. At other times, the next car will arrive before the cycle stock has been fully consumed, as shown during demand period D3.

The average inventory of that material over long periods of time will be:

$$\text{Inv (avg)} = \text{Safety Stock} + \tfrac{1}{2}\text{ Cycle Stock}$$

Because our raw material inventory consists of six materials, each with different arrival times, the total inventory at any time should approach the average, because the out-of-phase sawtooth waves will smooth each other out.

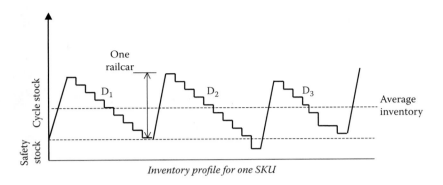

Inventory profile for one SKU

FIGURE A.1
Inventory profile for one raw material.

Because each material from Supplier 1 is received in 140,000-pound railcars, the cycle stock for each material will be 140,000. Four of the six materials from Supplier 1 are also supplied by Supplier 2. Those arrivals are probably not currently coordinated with arrivals from Supplier 1, one reason for the high raw material inventory. There is no reason the orders for a specific material from each supplier can't be coordinated in the future state, so that when it is time to reorder a material, it will be decided which of the two suppliers is to receive the order. If managed in that fashion, each material will require only the one railcar of cycle stock, regardless of supplier.

The specific methods for calculating safety stock are covered in Chapter 15; for now we will assume that the only significant risk is that railcars could be as much as two days late, so we will carry safety stock in the amount of two days worth of railcars. It is likely that the six raw materials are not consumed at equal rates, so the two days of safety stock will be different amounts of material for each of the six raw material types. But because the process consumes an average of 2.8 railcars of raw material each day, two days of safety stock will total 5.6 railcars, distributed among the raw materials in proportion to their daily consumption.

Thus, the total cycle stock will be:

6 raw material types × 140,000 lbs each = 840,000 lbs

The total safety stock will be:

2 days × 2.8 railcars/day × 140,000 lbs/railcar = 784,000 lbs

The average inventory will be:

$$784{,}000 \text{ lbs} + 1/2(840{,}000 \text{ lbs}) = 1{,}204{,}000 \text{ lbs}$$

Applying Little's law (*Days of supply* = *WIP* ÷ *Throughput*), with an average flow of 2.8 railcars, or 392,000 lbs, per day, the average inventory is:

$$1{,}204{,}000 \text{ lbs} \div 392{,}000 \text{ lbs/day} = 3.1 \text{ days}$$

Several important things should be noted:

- The safety stock determination used here is approximate; more refined methods are introduced in Chapter 15.
- A more accurate determination of the appropriate safety stock levels will require more detail on the specific raw material ordering process, and statistical data on historical arrival variation.
- More than half of the average inventory is safety stock. This is not unusual in cases with frequent arrivals, but high variability in arrival times.
- The rules of thumb often cited in the literature that safety stock should be some percentage of cycle stock, 10 percent, 20 percent, or some other pre-specified ratio, are among the poorest ways to estimate safety stock needs.
- If our sheet goods plant takes ownership of the materials when they leave the suppliers' facilities, our total raw material inventory includes approximately 7 days of transportation inventory. This is sometimes called "pipeline inventory" even when it is not transported by pipeline.

Appendix B

References

Abdullah, Fawaz. *Lean Manufacturing Tools and Techniques in the Process Industry with a Focus on Steel.* PhD dissertation, University of Pittsburgh, 2003.

Althaus, Ronald. "Silent seasonality." *APICS Magazine,* May/June 2008.

Arnold, J. R. Tony, Stephen N. Chapman. *Introduction to Materials Management.* Upper Saddle River, NJ: Pearson Prentice-Hall, 2004.

Bowersox, Donald J., David J. Closs. *Logistical Management.* New York: McGraw-Hill, 1996.

Chopra, Sunil, Peter Meindl. *Supply Chain Management—Strategy, Planning, & Operation.* Upper Saddle River, NJ: Pearson Prentice-Hall, 2007.

Cox, James F. III, John H. Blackstone, Jr. *APICS Dictionary,* 10th ed. Alexandria, VA: APICS, Educational Society for Resource Management, 2002.

Cunningham, Jean E., Orest Flume, Emily Adams. *Real Numbers: Management Accounting in a Lean Organization.* Durham, NC: Managing Times Press, 2003.

Ford, Henry. *Today and Tomorrow.* Portland, OR: Productivity Press, 1988.

George, Michael L. *Lean Six Sigma.* New York: McGraw-Hill, 2002.

Goldratt, Eliyahu M. *Theory of Constraints.* Great Barrington, MA: North River Press, 1990.

Goldratt, Eliyahu M., Jeff Cox. *The Goal.* Croton-on-Hudson, NY: North River Press, 1984.

Gooch, James, Michael George, Douglas Montgomery. *America Can Compete.* Dallas, TX: Institute of Business Technology, 1987.

Hopp, Wallace J., Mark L. Spearman. *Factory Physics.* New York: Irwin/McGraw-Hill, 2001.

Imai, Masaaki. *Gemba Kaizen—A Commonsense, Low-Cost Approach to Management.* New York: McGraw-Hill, 1997.

Invistics Corporation. *Processing Lean: Modifying Traditional Techniques for Complex Environments.* www.invistics.com, 2004.

Japan Institute of Plant Maintenance. *TPM for Every Operator.* Portland, OR: Productivity Press, 1996.

Kennedy, Ross, The Centre for TPM (Australasia). *Examining the Processes of RCM and TPM.* www.ctpm.org.au, January 2006.

Liker, Jeffrey K. *The Toyota Way.* New York: McGraw-Hill, 2004.

Maskell, Brian, Bruce Baggaley. *Practical Lean Accounting.* New York: Productivity Press, 2004.

Nakajima, Seiichi. *Introduction to TPM.* Cambridge, MA: Productivity Press, 1988.

Ohno, Taiichi. *Toyota Production System: Beyond Large Scale Production.* New York: Productivity Press, 1988.

Productivity Press Development Team. *Kaizen for the Shopfloor.* New York: Productivity Press, 2002.

Rath & Strong's Lean Pocket Guide. Lexington, MA: Rath & Strong, 2006.

Ross, Joel E., William C. Ross. *Japanese Quality Circles and Productivity.* Reston, VA: Reston Publishing Co, 1982.

Rother, Mike, John Shook. *Learning to See*. Cambridge, MA: Lean Enterprise Institute, 2003.

Rummler, Geary A., Alan P. Brache. *Improving Performance—How to Manage the White Space on the Organization Chart*. San Francisco, CA: Jossey-Bass, 1995.

Sayer, Natalie J., Bruce Williams. *Lean for Dummies*. Hoboken, NJ: Wiley, 2007.

Schonberger, Richard J. *World Class Manufacturing: The Lessons of Simplicity Applied*. New York: The Free Press, 1986.

Shingo, Shigeo. *Quick Changeover for Operators: The SMED System*. New York: Productivity Press, 1996.

Shingo, Shigeo, Andrew P. Dillon. *A Revolution in Manufacturing: The SMED System*. Cambridge, MA: Productivity Press, 1985.

Shunta, Joseph P. *Achieving World Class Manufacturing through Process Control*. Englewood Cliffs, NJ: Prentice-Hall, 1995.

Smalley, Art. *Creating Level Pull*. Brookline, MA: Lean Enterprise Institute, 2004.

Smith, Wayne K. *Time Out*. New York: John Wiley & Sons, 1998.

Teed, Nelson J. "Origins and reality of lean manufacturing." *APICS Magazine*, January 2001.

Umble, Michael, Mokshagundam L. Srikanth. *Synchronous Manufacturing, Principles for World Class Excellence*. Cincinnati, OH: South-western Publishing Co., 1990.

Womack, James P., Daniel T. Jones. *Lean Thinking*. New York: The Free Press, 2003.

Womack, James P., Daniel T. Jones, Daniel Roos. *The Machine That Changed the World*. New York: Macmillan, 1990.

Index

A

"A" type process
  implementation, 66
  material flow patterns, 28
  product differentiation points, 32
ABC classification, 296–297, 299
Abdullah, Fawaz, 181
Accidents, 45, 273
Adjustments in design, 224
Air freight costs, 291
Alternatives, safety stock, 290–291
Althaus, Ronald, 302
Alvarez, Juan Manuel, 266
Analysis
  bottlenecks, 91
  demand variability, 298–299
  flow, 91
  fundamentals, 89–90
  further opportunities, 92
  non-value-adding activities, 91
  questions checklist, 92–93
  variability, 92
  voice of the customer, 90
  waste, 90–91
Andons, 12, 135, 144, *see also* Kanban;
  Visual displays
Annual shutdowns, 13
Approximation, EOQ calculations,
  218–219, 224
*A Revolution in Manufacturing: The
  SMED System,* 124
AS/RS, *see* Automatic storage and
  retrieval system (AS/RS)
Assemblers, 20
Assembly processes and plants
  assemble-to-order strategy, 231
  cellular manufacturing, 178–180
  finish-to-order strategy, 230–231
  historical developments, 4–5
  material flow patterns, 26, 28
  non-value-added time, 78, 81

pre-cellular manufacturing, 177–178
process industry contrasts, 19–21,
  34–35
pull replenishment systems, 243–245
solutions, 203–205
timeline, 78, 81
work-in-process visibility, 26
Asset productivity, 23
Automated packaging lines, 41
Automatic storage and retrieval system
  (AS/RS), 42
Autonomation, 8, 41, *see also* Jidoka
Autonomous maintenance, 112
Available time model, 216–217

B

Bagging areas
  changeover, 127
  takt time, 62
  time on hand (waiting) waste, 41
Baka-yoke, 13
Basic producers, 20
Batches
  overproduction waste, 39
  process step data box, 72
  product transitions, 126
  push-pull interface, 247
  tank heels, 47
Benefits
  cellular arrangements, 194–195
  finish-to-order strategy, 233–234
  product wheels, 226–227
  total productive maintenance, 114
  value stream mapping, 58–59
Best practices, value stream mapping
  fundamentals, 88
  geographic arrangement, 84, 86
  logical flow, 84, 86
  parallel equipment, 81, 84
Bins, *see* Kanban

Material flow patterns, *see also* Flow
 process industries, distinguishing
  characteristics, 26, 28
 value stream mapping, 58, 77, 80,
  93–95, 97
Material inventory, raw, *see* Inventory;
 Raw materials inventory
Material Requirements Planning (MRP),
 78
Mathematical detail, *xxi–xxii*
Measures and metrics
 availability, 115
 equipment effectiveness, 115–116
 fundamentals, 114
 inappropriate use, 302–303
 kaizen events, 159
 OEE, 118–121
 performance, 115–116
 quality, 116
 UPtime, 116–118, 120–121
 visual management, 135, 144–145
 VSM data boxes, 120–121
Meindl studies, 229, 285, 290
Mismanagement, 76
Mix, leveling, 204, 209, *see also* Heijunka
Models
 available time model, 216–217
 computer simulation, bottlenecks, 168
 continuous review, 279–280
 estimate economic optimum, 217–219
 fixed interval replenishment, 275–277,
  288
 fixed order interval, 277
 fixed quantity replenishment, 278–280
Movement
 bottlenecks, 165–167
 waste, 37, 47–48, 53, 91
MTO, *see* Make-to-order (MTO) strategy
MTS, *see* Make-to-stock (MTS) strategy

N

Nakajima, Seiichi, 121
Natural ingredients, 49
Near-bottlenecks, 168, 171, *see also*
 Bottlenecks
Necessary waste, 51–52
Next quarter's sales, 301–302

Non-value-adding (NVA) activities
 analysis of, 91
 timeline, 78, 81
 value stream mapping, 58

O

Obsolete inventory, 274, *see also* Inventory
Off-spec materials, 44, 51
Ohno, Taiichi
 autonomation, 40
 categories of waste, 37–38, 90–91, 201
 cell productivity, 179
 challenges in process industries, 21
 defective parts waste, 48
 employee empowerment, 295
 five whys, 13, 101
 jidoka, 12
 just-in-time, 9, 15
 kanban, 244
 labor productivity, 22
 manufacturing advancement, 6
 movement waste, 47
 production rate, 203
 production rate variability reduction,
  10
 pull, 241–244
 supermarket inspiration, 269
 time added to process through waste,
  50–51
 TPS basis, 8
 visual management, 137
 wastes, 8, 50–52
Operating rules, 199
Operational stability, *see* Heijunka
Operations and business integration, 300
Optimum sequence determination, *See*
 sequence
Orders, mismanagement, 76
Outage, *see* Shutdowns
Overall equipment effectiveness (OEE)
 availability, 115
 bottlenecks, 164
 calculation, 118–120
 capital *vs.* labor, 23
 cycles, 62, 70
 fundamentals, 114
 material flow, 97

About the Author

Peter L. King is the president of Lean Dynamics, LLC, a manufacturing improvement consulting firm located in Newark, Delaware. Prior to founding Lean Dynamics, Pete spent 42 years with the DuPont Company, in a variety of control systems, manufacturing systems engineering, continuous flow manufacturing, and lean manufacturing assignments. The past 18 years have been spent applying world class manufacturing techniques to a wide variety of products, including sheet goods like DuPont™ Tyvek®, Sontara®, and Mylar®; fibers such as nylon, Dacron®, Lycra®, and Kevlar®; automotive paints; performance lubricants; bulk chemicals; adhesives; electronic circuit board substrates; and biological materials used in human surgery. On behalf of DuPont, Pete has consulted with key customers in the processed food and carpet industries. Pete retired from DuPont in 2007, leaving a position as Principal Consultant in the Lean Center of Competency.

Pete received a bachelor's degree in electrical engineering from Virginia Tech in 1965. He is Six Sigma Green Belt certified (DuPont, 2001) and Lean Manufacturing certified (University of Michigan, 2002). He is a member of the Association for Manufacturing Excellence, APICS, and the Society of Manufacturing Engineers. He is active in the Institute of Industrial Engineers, and is currently President of the Process Industries Division.

DuPont™, Tyvek®, Sontara®, Kevlar® are trademarks or registered trademarks of E. I. D duPont de Nemours and Company. Mylar® is a trademark of DuPont Teijin Films; Dacron® and Lycra® are trademarks of Koch.